Garbi

Because the author
chose an Argonne
design for the
advanced reactor,
so that you should
have this

Chuck

Signed up for a cruise
along the inland
waterway next March,
so I may not come to
SD as planned

Whittle BOOKS

NUCLEAR RENEWAL

ALSO BY RICHARD RHODES

Verity

A Hole in the World: An American Boyhood

Farm: A Year in the Life of an American Farmer

The Making of the Atomic Bomb

Looking for America: A Writer's Odyssey

The Ozarks

The Inland Ground:
An Evocation of the American Middle West

Fiction

Sons of Earth

The Last Safari

Holy Secrets

The Ungodly

NUCLEAR RENEWAL

COMMON SENSE ABOUT ENERGY

Richard Rhodes

WHITTLE BOOKS

IN ASSOCIATION WITH

VIKING

VIKING
Published by the Penguin Group
Penguin Books USA Inc., 375 Hudson Street, New York,
New York 10014, U.S.A.
Penguin Books Ltd, 27 Wrights Lane, London W8 5TZ, England
Penguin Books Australia Ltd, Ringwood, Victoria, Australia
Penguin Books Canada Ltd, 10 Alcorn Avenue,
Toronto, Ontario, Canada M4V 3B2
Penguin Books (N.Z.) Ltd, 182–190 Wairau Road,
Auckland 10, New Zealand

Penguin Books Ltd, Registered Offices: Harmondsworth,
Middlesex, England

First published in 1993 by Viking Penguin,
a division of Penguin Books USA Inc.

10 9 8 7 6 5 4 3 2 1

Grateful acknowledgment is made for permission
to reprint excerpts from the following copyrighted works:
The Nuclear Energy Option by Bernard L. Cohen.
By permission of Plenum Publishing Corp.
Nuclear Politics by James M. Jasper. Copyright © 1990 by
Princeton University Press. Used with permission.

LIBRARY OF CONGRESS CATALOGING-IN-PUBLICATION DATA
Rhodes, Richard.
Nuclear renewal: common sense about energy/Richard Rhodes.
p.   cm.
ISBN 0-670-85207-4
1. Nuclear industry—United States. I. Title.
HD9698. U52R46 1993
333. 792'4'0973—dc20   93-22021

Printed in the United States of America
Set in Sabon
Designed by Kathryn Parise

C O N T E N T S

NUCLEAR
RENEWAL

CHAPTER 1

A FAILURE OF
MANAGEMENT

Nuclear power isn't dead. It runs France and will soon run Japan. The United States relies on it, with some 110 commercial reactors generating about one-fifth of the nation's electricity, more reactors than any other country in the world. Luddites and small-worlders have demonized nuclear power. Its promoters and managers, talented at shooting themselves in the foot, have squandered its original fund of public goodwill. Despite these political liabilities, electricity from nuclear fission continues to be the most comprehensive source of energy available to meet growing U.S. demand—the cleanest and the safest of major sources. Some people will find these statements scandalous. Others will welcome them as nothing less than common sense and plain truth.

The purpose of this short book is straightforward: to review what happened to nuclear power to bring it to its present impasse in the United States, to report on its situation today, and to consider what its future might be. The subject is important, one about which citizens, especially business and political leaders, ought to be better informed. Energy demand is increasing in the United States. It will continue to increase as our population grows. It's also shifting significantly to electricity, just as it is in countries like France, where high energy costs have always enforced stringent conservation. If nuclear power is the cleanest and safest significant form of energy available, as a majority of competent scientists and engineers conclude, then it ought to be supported, not condemned.

"The failure of the U.S. nuclear power program," *Forbes* magazine announced in a ferocious cover story in 1985, "ranks as the largest managerial disaster in business history." *Forbes* liberally dispenses blame across the entire program before specifying who might be primarily responsible; its list of miscreants is a useful back-of-the-cuff summary.

The truth is that nuclear power was killed, not by its enemies, but by its friends:

- The federal government and the Nuclear Regulatory Commission, which not only botched the day-to-day management of the program but also failed to consider the economic cost of the regulations it imposed.

- The equipment manufacturers, who maintained that nuclear power was just another way of boiling water.

- The contractors and subcontractors, the designers and engineers and construction managers who, insulated by their own cost-plus contracts, had little incentive to question the cost-effectiveness of the NRC's dictates.

- The utility executives, who believed that no matter what happened to cost and construction schedules, the rate commissions would somehow provide the revenues to bail them out.

- And the state regulatory commissions themselves, whose grossly inadequate oversight of the schemes, ambitions, and monstrous expenditures for nuclear projects made it easier for all of the above to betray consumers and investors alike.

On the evidence, though, *Forbes* finds utilities management most culpable, a culpability it demonstrates by running the numbers on thirty-five nuclear power plants that had not yet gone on line in 1984. They all experienced delays and cost overruns, *Forbes* notes, but "for all their common technical, social, and political environment, the costs of these plants differ widely, ranging from a commendable $932 a kilowatt for Duke Power's McGuire 2 station to a grotesque $5,192 a kilowatt for Long Island Lighting's Shoreham plant. . . . The disparities in cost are so great as to make a prima facie case for mismanagement in the first degree."

Forbes's eagerness to blame obscures a less sensational but more valuable lesson its data disclose: that some utilities did better than others. Lay observers have wondered for years why the U.S. nuclear power business has stalled while the French and Japanese industries, built on U.S.-designed reactor systems, have thrived. There are several answers to that question (in a later chapter I'll take them up). One answer is simply historical accident: the U.S. industry is various and private, not unified and central, and different utilities operate at different levels of skill. Certainly some U.S. utilities botched the job. Others—Yankee Atomic Electric, Duke Power, Commonwealth Edison of Chicago—built as

cost-effectively and manage as well as the French

and the Japanese have done. "Some of the utilities that jumped on the nuclear bandwagon were ready to take on nuclear responsibilities," Nuclear Regulatory Commissioner Victor Gilinsky told an industry audience in 1982, "but some were not." Unfortunately, Gilinsky added, the nuclear power business "is hostage to the worst performers in the industry."

Not surprisingly, the "worst performers" have passed along the blame they've received. They've blamed interest rates, the antinuclear movement, "regulatory ratcheting," public hostility, even just plain bad luck (which is usually spelled *TMI,* i.e., Three Mile Island). All these factors played a part. But the disparity in cost between the worst and the best performances, as *Forbes* properly emphasizes, proves they weren't determining. The quality of management was.

A 1981 study of power-plant cost escalation found that interest payments totaled not more than 10 percent of the cost of plants built up to 1978. After that, interest costs went up, but because Three Mile Island led to expensive mandated design changes, so did total costs. "Interest charges," writes the sociologist James M. Jasper, "were a significant but derivative cause of nuclear energy's high cost. They arose from the enormous construction delays that plagued the industry, and from the increasing skepticism of the investment community."

Beginning around 1975, antinuclear protesters descended on nuclear power. Their assault drew major attention on television and in the press. Since nuclear power has stalled, they appear to have succeeded in their campaign. But the real story is otherwise—sadder and also more instructive.

Antinuclear activists, like the ecological activists who have succeeded them, tended to be better educated than the national average, younger rather than older, women rather than men, middle-class rather than working-class, white-collar rather than blue. They were more likely to work in the service sector than on the assembly line. Someone else picked their cotton, built their cars, and pumped their sewage. Their style was moralist and elitist, a secular borrowing from religion. They were at least as guilty of technological enthusiasm unsupported by evidence and experience as the early promoters of nuclear power, except that their favorite technologies happened to be solar, wind, hydro, and geothermal. (Each of which has its own significant environmental consequences, as demon-

strated by Hydro-Quebec's vast hydroelectric

scheme under development at James Bay. That construction project is one of the largest ever undertaken; it will alter an area the size of France.)

But of all the possible industrial systems these ecological moralists might have attacked, why did they choose one so relatively benign as nuclear power? No one has been killed in a U.S. commercial nuclear power accident in three decades of successful operation, nor has commercial power released more than minimal amounts of radioactivity. Why not target the chemical industry, with its vast inventory of toxic wastes? Why not automobiles, with their ubiquitous and unhealthy smog and their 50,000 deaths a year? For that matter, why not coal power, which produces evident damage from acid rain and kills about 30,000 people per year with air pollution?

The generation that came of age in the 1960s and 1970s was the generation of "duck and cover," men and women who as children during the 1950s learned to duck under their school desks and cover their heads to practice protecting themselves from nuclear attack, who worried that fallout from atmospheric nuclear testing would poison their milk, whose parents dug fallout shelters in their backyards. In this way, as in its early and premature commercialization, nuclear power was a casualty of the Cold War.

Since its discovery at the turn of the century, radiation has been credited with unusual power, power associated with the concept of transmutation, of passage through destruction to rebirth. Businessmen tend not to take such symbolic constructs seriously, but perhaps they should educate themselves to their influence: what people believe affects profoundly how they think and feel.

The mines in the Erzgebirge of Germany and Czechoslovakia, the world's first source of uranium, supplied the pitchblende from which Marie and Pierre Curie first precipitated radium in 1898. When doctors around the world began using radium to treat cancer, Erzgebirge spas such as Carlsbad and Marienbad began advertising the tonic radioactivity of their natural spring waters. A German firm successfully marketed a radioactive toothpaste in those innocent early days. Unwitting overexposure to radiation from medical X rays and in industrial processes, such as the hand-painting of radium-dial watches, first revealed radiation's health risks. Hiroshima and Nagasaki and then the accidental exposure of Marshall Islanders and Japanese fishermen to

fallout from nuclear-weapons testing convinced

millions that radiation was uniquely dangerous. (Significantly, demanding that the superpowers move their nuclear-weapons testing underground was the only successful grass-roots arms-control initiative in the entire history of the arms race.) Radiation seems all the more sinister because our species has evolved no sensory organs that can detect it even at lethal doses. "Nuclear energy was conceived in secrecy," writes one student of public attitudes, "born in war, and first revealed to the world in horror. No matter how much proponents try to separate the peaceful from the weapons atom, the connection is firmly embedded in the minds of the public."

Although it serves many useful purposes, including lifesaving medical treatment, people have come to associate radiation with violence and darkness and death. "Even before the accident at Three Mile Island," a 1991 review of opinion studies in the journal *Science* concludes, "people expected nuclear reactor accidents to lead to disasters of immense proportions. When asked to describe the consequences of a 'typical reactor accident,' people's scenarios were found to resemble scenarios of the aftermath of nuclear war." Burying nuclear wastes fused with glass in rock formations a mile underground would isolate them safely and effectively for at least tens of thousands of years, but four surveys cited in *Science* of the words people associate with a nuclear waste repository found that negative consequences and concepts predominate: *danger, dangerous, unsafe, disaster, hazardous, poisonous,* and similar negatives were the most common word associations. Positive imagery was rare, only 1 percent of the total.

These baleful associations pervade public attitudes toward nuclear power, but they still don't fully explain the motives of the antinuclear activists. David E. Lilienthal, the first chairman of the U.S. Atomic Energy Commission, a wise and thoughtful man, identifies a possibly unconscious purpose in his book *Atomic Energy: A New Start*, published in 1980 not long before he died. Lilienthal suspected antinuclear activists might be attacking nuclear power as a surrogate for nuclear weapons:

> It is as if some nuclear opponents believe that abolishing the lesser evil (radiation from malfunctioning nuclear plants) will somehow reduce the greater evil (the bomb). This exercise in futility illustrates how hopeless and helpless most people feel about the clear and growing danger of atomic war. The nuclear arms race has gotten so far out

of control that, for many, it seems quite impossible to confront it with any emotion but despair or resignation. Atomic electricity, on the other hand, is not yet beyond the reach of corrective public action. As a surrogate for the bombs, atomic energy can be fought, protested, and possibly defeated, even though such a victory would not resolve the issue of the presence of atomic weapons.

Such a displacement of protest may be understandable, but it also represents a failure of moral courage. If activists were concerned about nuclear war, they should have picketed the air bases and missile silos where the weapons were stored or the factories and embassies of the countries that built them, not nuclear power plants. The truth is, attacking nuclear power was easier and safer than attacking the military establishment. No one risked losing his job or being branded a traitor or going to jail for anything more serious than trespassing.

Despite its high visibility, the rise of nuclear activism was far less destructive to the development of nuclear power than was the 1973 Arab oil embargo, which ended the long, steady progress of energy demand in the United States and changed the basic assumptions of the utility industry and the supporting investment community. Utilities subsequently canceled orders for some 100 nuclear power plants—but, significantly, they also canceled orders for some 70 coal-fired plants. "At the time that the energy crisis went into remission," Bertram Wolfe, the head of nuclear power at General Electric, told an MIT conference audience in 1990, "the economy was pulling out of a period of stagnation, and energy supplies were plentiful. . . . We canceled a number of plants and still had a glut of capacity. With oil prices plummeting and with this glut of generating capacity, one could hardly expect public enthusiasm for energy development of any type." Capacity continued at surplus, Wolfe noted, until about 1989: "So you could be against coal plants or nuclear plants or gas pipelines or new dams, you could be against picking up wood in the forest, you could be against everything and it did not matter."

Nor was activist intervention broadly successful even on its own terms. Delays in licensing nuclear power plants were dwarfed by management- and regulatory-induced construction delays. In one Federal Power Commission study of 23 plants, changes in regulatory requirements and legal challenges caused 32 months of delay, but strikes, labor shortages, late delivery of major equipment, component failures, and other essentially managerial problems caused 220 months of delay. James Jasper summarizes:

In the last twenty years no nuclear reactor in the United States has been stopped because of its inability to obtain a license. A handful of construction permits have been temporarily denied or delayed, but these have typically resulted in design changes rather than project cancellations. One may have to look back to Bodega Head and similar cases in the 1960s to find proposed reactors abandoned because of expectations that construction permits would be denied. It was twenty years later (1984) that, for the first time, an operating permit was denied to a completed plant, for Byron 1 in Illinois, but it was granted after further hearings.

Activism certainly contributed to New York State's decision to shut down Shoreham Power Station on Long Island in 1989. But even that plant had previously been awarded a full-power license by the Nuclear Regulatory Commission. Almost everywhere, nuclear activists jumped on a bandwagon that had already broken loose and was careening downhill.

"Regulatory ratcheting" has also been a factor in stalling nuclear power. The phrase refers to the NRC practice of piling on new regulations without reviewing the old. "The NRC did not withdraw requirements made in the early days on the basis of minimal experience," writes radiation physicist Bernard Cohen of the University of Pittsburgh, "when later experience demonstrated that they were unnecessarily stringent. Regulations were only tightened, never loosened." Regulatory ratcheting directly delayed construction and operation; more subtly, it may have contributed to poor management as well.

Cohen questions whether the piling on of regulatory requirements was "necessary, justified, or cost-effective." The answer to that question, he argues, depends on how one defines those terms:

> The nuclear regulators of 1967 to 1973 were quite satisfied that plants completed and licensed at that time were adequately safe, and the great majority of knowledgeable scientists agreed with them. With the exception of improvements instigated by lessons learned in the Three Mile Island accident, which increased the cost by only a few percent, there were no new technical developments indicat-

ing that more expenditures for safety were needed. In fact, the more recent developments suggested the contrary. . . . Perhaps the most significant result of safety research in the late 1970s was finding that the emergency core cooling [system] works better than expected and far better than indicated by the pessimistic estimates of nuclear power opponents. . . .

Clearly, the regulatory ratcheting was driven not by new scientific or technological information, but by public concern and the political pressure it generated. Changing regulations as new information becomes available is a normal process, but it would normally work both ways. The ratcheting effect, only making changes in one direction, was an abnormal aspect of regulatory practice unjustified from a scientific point of view. It was a strictly *political* phenomenon. . . .

If regulation was excessive, it may also have been too legalistic. "Having a set of precise standards to be met," writes Jasper, "indicated to many utilities that they *only* had to meet those standards." A 1984 NRC study, Jasper reports, "found that a key difference between utilities that succeeded with nuclear energy and those that didn't was that the former treated regulatory guidelines as a bare minimum rather than a satisfactory goal."

Significantly, the U.S. regulatory climate is distinctly adversarial. In countries where nuclear power has found greater success, regulators work directly with utilities to achieve safe, reliable operation. By contrast, U.S. utilities can be fined for problems they voluntarily report. One of the horror stories I heard while interviewing managers and engineers for this book was the dark comedy of second-guessing that led to the partial meltdown in 1966 of the Fermi 1 experimental breeder reactor built with private funds in the 1950s and '60s near Detroit.

A breeder reactor transmutes an extra charge of uranium to plutonium in the course of operation, as well as generating heat. "Two subassemblies partially melted," Don Wille, a nuclear engineer, told me. Wille is a senior project manager at Stone & Webster, the engineering construction firm. He was running the Detroit office of the Atomic Energy Commission at the time Fermi 1 was built. As Wille recalls it :

Fermi 1 was a sodium-cooled reactor. The inlet-flow sodium came into the bottom of the reactor vessel, below the core, and flowed up and through and around the fuel assembly. Because of the way the pipes came into the vessel, the designers felt they needed a flow diverter, a truncated cone called a pedestal, to promote the smoother flow of sodium fluid. Well, that was fine. But then, when they were about to go into AEC regulatory hearings, they saw the possibility in the event of a meltdown that the diverter could allow molten uranium to collect at the bottom of the vessel and burn through. What if the regulators saw the same possibility? they asked themselves. So they decided on the spur of the moment that the design also needed a system of deflectors.

They took the problem to a local shop and had zirconium plates made up, but the plates weren't substantial. They were thin zirconium stock. They had them drilled with holes and they mounted them with small screws instead of heavy bolts. Spirited this thing back to the plant one night and installed it. They didn't have drawings, they didn't have calculations, they didn't have procedures. They were trying to anticipate what the regulators might say. And the result was catastrophic. The plates eventually came loose, migrated, blocked 95 percent of the sodium flow. That's when the fuel assemblies melted.

Fermi 1 was an utter ruin, and the disaster effectively foreclosed the development of a commercial breeder reactor in the United States for the next thirty years.

But Don Wille didn't tell me that horror story only to blame the sour regulatory climate in the United States. His point was also that what the Fermi 1 designers did was a failure of quality assurance—of management. "Did the utilities that bought the plants really understand that this was serious technology?" I had asked Wille earlier in our interview. "Only a few did," he answered. "Knowing about something isn't the same as being steeped in the culture. Some developed it. But we have 110 reactors and we probably have 50 utilities operating them. And if management doesn't pay attention, then you end up with people sleeping at the controls."

UNLOCKING A GIANT

Two historic tests frame the development of nuclear power and the possibility of its revitaliza-
tion, events separated by forty-four years. One
test, the earlier, conducted in Chicago in 1942,
marked the beginning of a new age—a "new
world," a participant called it, and it was. The
other test, the more recent, conducted in Idaho
in 1986, marked a point of renewal—a hopeful
new beginning. Between the two events runs the
remarkable story of the deployment of a new
source of energy comparable in its potential to
oil or gas or coal, a deployment measured not
in centuries but in decades, a deployment with
fateful consequences for national security as well

as commercial growth. "There are overtones in this development," the physicist and statesman J. Robert Oppenheimer commented in 1957, "that have been absent in power developments . . . such as the diesel engine and the gas turbine: overtones of pride and terror, of mystery and hope."

The first historic test—of the world's first man-made nuclear reactor, Enrico Fermi's famous "pile," CP-1—took place on December 2, 1942, the second day of wartime gasoline rationing, in a squash court under the stands of the University of Chicago's Stagg Field. Fermi, the Italian Nobel laureate physicist, led the team that day. An eyewitness remembers the scene:

> It was terribly cold—below zero. . . . Inside the hall of the west stands, it was as cold as outside. We put on the usual gray (now black with graphite) laboratory coats and entered the doubles squash court containing the looming pile . . . and then went up on the spectators' balcony. The balcony was originally meant for people to watch squash players, but now it was filled with control equipment and readout circuits glowing and winking and radiating some gratefully received heat.

The pile, as it waited in the dark cold of Chicago winter, was a black, greasy, flattened-ovoid hulk—a doorknob as big as a garage—stacked with 771,000 pounds of 16-inch graphite bricks, 80,590 pounds of uranium-oxide pucks, and 12,400 pounds of uranium-metal slugs, the uranium components dropped into blind holes bored into the graphite bricks in a roughly spherical lattice. CP-1 cost about $1 million to build. Its only visible moving parts were its various control rods. It had no shielding. It had no cooling. Fermi intended to run it no hotter than half a watt, hardly enough energy to light a flashlight bulb. But no one doubted that its mechanism, if it worked, could someday be applied to the production of power. Such power would keep submarines in perpetual motion underwater, a few people in the U.S. Navy had quickly realized. Others, including Fermi's young physicist colleague Walter Zinn, were already thinking about power for civilian electricity.

Accustomed as we are to color televisions and computers, it's difficult to conceptualize the workings of a machine made of components as primitive as hand-hewn graphite bricks and chunks of heavy metal. CP-1 was a sophisticated mechanism, but its sophistication operated on a scale not of squash courts but of atoms.

A uranium atom is as large an assembly of atomic particles—protons and neutrons—as can naturally exist. A fundamental force of nature that physicists call the strong force holds the nucleus together, but the 92 protons in the uranium nucleus repel each other electrically with almost enough push to cancel out the cohesion. This near balance between the two forces makes a uranium atom wobbly and unstable, like a water-filled balloon. Inject one more nuclear particle—a neutron slips in easily—and a uranium nucleus will wobble so wildly that it will sometimes actually pull apart and divide in two. Once a nucleus has "fissioned" in this way, the strong force reasserts itself within the separate pieces, reorganizing them into smaller atoms. The two new pieces, both positively charged, then repel each other powerfully. They fly apart. The energy of their repulsion manifests itself as heat. And that heat, from collective trillions of fissioning uranium atoms, can be used to heat water to steam to make electricity. (Bombs also produce heat, but by significantly different arrangements.)

Within a fraction of a second after fission, the two new smaller atoms eject the extra neutrons they now contain—on average, slightly more

than two per original uranium atom. Those secondary neutrons can be made to encounter other
uranium nuclei in turn and fission them.

Different kinds of uranium have different
properties. In nature they come mixed together,
and since they're chemically identical, they're extremely difficult to separate. The kind that's important to nuclear power production and to
bomb-making is called U-235. The other kind,
U-238, is much more common. The problem
with U-238 is that it captures fast neutrons, including the ones that a fissioning U-235 atom
ejects. Having done so, it doesn't fission and it
doesn't release a burst of heat. It transmutes to
plutonium. So U-238 is useful for making plutonium, but it gets in the way of power production.

Fortunately for nuclear power, U-238 is im-
pervious to neutrons that have been slowed
down to room temperature—slow neutrons, as
they're called. The arrangement Fermi and his
crew were testing that cold winter day was designed to slow down the secondary neutrons released in the fission of U-235. The slowed
neutrons could then avoid capture by U-238 and
fission more U-235 atoms instead. That's what
the carefully spaced arrangement of graphite and
uranium was about. Neutrons can be slowed by
colliding them like billiard balls against the nu-

clei of light elements such as hydrogen, helium, sodium, or carbon. The graphite bricks of CP-1, embedded with hundreds of pucks and slugs of uranium, supplied a field of carbon atoms all around the uranium to slow—to "moderate"— the secondary neutrons.

When enough neutrons fissioned the U-235 nuclei, Fermi reasoned, a chain reaction should occur, each fission causing two more fissions, two causing four, four causing eight, eight causing sixteen, in a geometric progression that could ultimately generate enough heat and radiation to burn up the pile if Fermi didn't limit the reaction with control rods. In CP-1 the handmade wooden rods were wrapped with sheets of cadmium, a metal that hungrily absorbs neutrons. By moving one or more control rods in or out of slots in the pile, allowing the cadmium to absorb greater or lesser numbers of neutrons, Fermi could accelerate, slow, or stop the chain reaction. (He began the experiment that morning with all the control rods in.) If something happened to the control rods, he had a suicide squad in reserve: three young scientists with jugs of cadmium-sulfate solution waited near the ceiling of the squash court, ready to flood the pile with cadmium and quench any runaway reaction at the risk of their lives.

Through the morning and early afternoon the historic experiment proceeded. (A man of regular habits, Fermi stopped for lunch.) Slowly the Italian physicist ordered the last control rod inched out of the hulking pile while he and his team checked the neutron counters and watched the recording drum that registered the pile's reactivity. Midafternoon, an eyewitness remembers, "suddenly Fermi raised his hand. 'The pile has gone critical,' he announced. No one present had any doubt about it." The reaction had become self-sustaining. The pile's neutron intensity at that point was doubling every two minutes as the chain reaction proceeded. Left uncontrolled for an hour and a half, that rate of increase would have carried CP-1 to a million kilowatts.

Fermi ran the pile for four and a half minutes at one-half watt before he shut it down. It was 3:53 p.m. "The Italian navigator has landed in the new world," an administrator whispered into the telephone to Washington. But émigré Hungarian physicist Eugene Wigner, another eyewitness, recalled long afterward how ambivalent about their achievement the experimenters felt. A force of nature had been released by an inspired application of human ingenuity; for good and for ill, forever after, it would have to be reckoned with:

Nothing very spectacular had happened. Nothing had moved and the pile itself had given no sound. Nevertheless, when the rods were pushed back in and the clicking died down, we suddenly experienced a let-down feeling, for all of us understood the language of the counter. . . . For some time we had known that we were about to unlock a giant; still, we could not escape an eerie feeling when we knew we had actually done it. We felt as, I presume, everyone feels who has done something that he knows will have very far-reaching consequences which he cannot foresee.

CP-1 was the immediate forerunner of massive uranium-graphite reactors that Du Pont built for the U.S. Army at Hanford, Washington, near Richland, later in World War II. They were intended for the production of plutonium for the first atomic bombs. In wartime their considerable heat was nothing more than a waste by-product; to dump it, Du Pont simply ran treated water from the Columbia River through the reactors; they heated the mighty Columbia by half a degree downstream.

The Army assembled a committee of experts in 1944 to think about postwar uses of atomic energy. The Tolman Committee was pessimistic about the atom's commercial prospects. "The development of fission piles solely for the production of power for ordinary commercial use," it concluded in its report, "does not appear economically sound nor advisable from the point of view of preserving national resources." Everyone involved in the early years of the new atomic energy enterprise believed that high-quality uranium ore was rare in the world, too rare to be diverted from weapons production. The Army even tried to buy up the entire known world supply. (About 6,000 tons of proven world reserves were accessible to the United States at the time; in comparison, a large modern reactor today requires a lifetime charge of some 4,000 tons.) Only at the end of the decade, when the federal government offered bonuses to uranium prospectors for high-quality finds, did the extensive resources of the Colorado Plateau come to light. Immediately after the war and well into the 1950s, almost the entire U.S. production of uranium and plutonium was dedicated to weapons.

The Atomic Energy Act of 1946 made atomic energy in all its manifestations an absolute monopoly of the federal government. All discoveries concerning atomic energy were to be considered "born" secret—treated as secret until formally declassified—and the penalty for divulging atomic secrets was life imprisonment or death. All uranium and plutonium became the property of the government, as beached whales once became the property of kings. No one could build or operate a reactor except under government contract, nor could one be privately owned. Proponents of the Atomic Energy Act had argued that atomic energy was too important to be left in the hands of the military. Apparently it was also too important to be conveyed into the hands of civilians.

Power enthusiasts in Congress grew restless. Cheap electricity from the atom was something they could send home. In 1948 the Congressional Joint Committee on Atomic Energy (JCAE), slowly becoming aware of its unusually broad mandate not only to propose legislation but also to monitor and control nuclear development, held extended hearings on the Atomic Energy Commission's program and wasn't happy with what it found. "The Joint Committee believes," it complained, "that reactor development should proceed with all possible speed, and disappointment therefore follows from the reflection that, in two and a half years [since the creation of the AEC], the commission has not broken ground on a single new-type high-power reactor."

Early in 1949 the AEC responded by establishing a division of reactor development. It began carving out the National Reactor Testing Station in the barrens of southeastern Idaho, where it authorized the building of an experimental breeder reactor. It authorized a materials-testing reactor there as well to study the effects of radiation on the materials from which reactors might be built, a step necessary to large-scale power production. Most significantly, and out of character with these other experimental projects, it directed Westinghouse in Pittsburgh to begin work on a power reactor for submarines.

Soon after CP-1's first operation in the middle of the city of Chicago, the experimental reactor had been dismantled and rebuilt with shielding and cooling twelve miles west in the Argonne Forest, near what would become the Argonne National Laboratory. At the Argonne site, Enrico Fermi, Walter Zinn, and their colleagues began building other research reactors. "In the '40s and '50s," says Charles Till, a vigorous Canadian who is Argonne's associate laboratory director for engineering research, "Argonne was a highly creative place. All sorts of different reactor designs were developed here in those days." The Illinois laboratory was designated as the AEC's center for commercial-reactor development in 1948. Walter Zinn, who had become director of Argonne in 1946, had recognized from the beginning that the real long-term promise of nuclear power derived from its potential for breeding. If U-235 was relatively scarce in the world, U-238 was plentiful; developing a commercial reactor that bred plutonium from U-238 would assure the United States of an essentially unlimited energy supply. Zinn had begun designing a breeder reactor even before the end of the war, and Argonne pursued the project vigorously afterward.

Another, and competing, center of reactor design developed within the U.S. Navy. The instigator of the Submarine Thermal Reactor (STR) project, which the AEC authorized Westinghouse to pursue, was Hyman Rickover. Born in Russia in 1900, Rickover was an engineering officer who had already produced a mine-sweeping system for the Navy and supervised the development of sonar. Busy making bombs, the AEC had wanted no immediate part in building nuclear submarines. Nine months of Rickover's carefully orchestrated prodding moved the commission to relent.

Zinn and Rickover had fought over the submarine reactor and Rickover had won. "It was mostly a turf fight," says Charles Till. "The designs weren't that different." One significant difference was the reactor's proposed coolant. By 1949 nuclear scientists and engineers had studied a number of different reactor configurations: cores of natural uranium, of enriched uranium (enriched in U-235), and of plutonium, in the form of metals and oxides as well as in solution; moderators including graphite, heavy water, light water, and paraffin; unmoderated reactors; coolants including water, air, helium, carbon dioxide, liquid sodium, and oil. Someone at the AEC joked that the only moderator they hadn't tried was sawdust, and the only coolant they hadn't tried was beer.

Zinn favored sodium, a metal that liquefies at 208 degrees Fahrenheit and has low neutron-absorbing properties as well as fifty times the thermal conductivity of water. Sodium happens to react violently with water and burn spontaneously on exposure to air. Rickover, an old submarine man, could foresee only disaster if sodium got loose in a submarine. He preferred water. Engineers were used to working with water, and Rickover was determined to get his submarines built.

Westinghouse was initially awarded $830,000 to begin designing the submarine system to

Rickover's specifications. One of his assistants, Lieutenant Commander Louis H. Roddis, had proposed the system Rickover wanted built. It was based on ideas developed by physicist Alvin M. Weinberg (who later became director of the Oak Ridge National Laboratory) in conversations with Fermi. The reactor Roddis proposed would consist of plates of enriched uranium, moderated and cooled by ordinary "light" water (so-called to distinguish it from "heavy" water, also used in reactor design, which incorporates a heavier form of hydrogen). Water would circulate among the uranium plates and then through a heat exchanger. A secondary water system, kept physically separate from the primary loop to isolate the primary's radioactivity, would pick up heat from the heat-exchanger and carry it to turbines to make steam. Water in the primary loop would be pressurized to prevent it from boiling. The complete mechanism would be reasonably compact, compact enough to fit into the hull of a new submarine, one much larger than the diesel submarines of World War II. It would require no oxygen to operate; with carbon-dioxide scrubbers and an oxygen generator for the crew, the revolutionary new submarine would be able to cruise for months without resurfacing or refueling.

In August 1951, the Navy issued a contract to the Electric Boat Company of Groton, Connecticut, for construction of a submarine powered by a pressurized-water reactor (PWR). At a public ceremony on June 14, 1952, President Harry S. Truman laid the keel, remarking that "widespread use of atomic power is still years away." The "paradox," Truman added, was that "most of our progress toward the peaceful application of atomic energy has come under the pressure of military necessity." On January 21, 1954, after a change of administrations, First Lady Mamie Eisenhower launched the nuclear-powered *Nautilus* with a bottle of champagne. Rickover was on hand, wearing new admiral's boards. "The Navy's project is the basis of the country's infant atomic industry," said a congressman who had fought the Navy for Rickover's promotion. "When civilian power comes, it will be a by-product of the Navy's work."

By then civilian power was on its way, at least at demonstration scale. Besides building nuclear submarines, the Navy was also sponsoring the design at Westinghouse of a larger light-water PWR for aircraft carriers. When President Dwight Eisenhower's budget men slashed the AEC's Navy, Air Force, and civilian reactor plans from their budget proposals in early 1953, the AEC and the JCAE persuaded the House Appropriations Committee to restore some sort of nuclear power project. "Now is the time," the AEC asserted in Congressional hearings, "to announce a positive policy designed to recognize the development of economic nuclear power as a national objective. . . . It would be a major setback to the position of this country in the world to allow its present leadership in nuclear power development to pass out of its hands." Congress responded cautiously by authorizing the AEC to spend up to $7 million to develop a land-based version of the aircraft-carrier PWR.

It's clear in hindsight that the careful evolutionary development of nuclear power in the United States, including the nurturing of a solid political constituency that might have sustained it through its subsequent troubles, was a casualty of the Cold War. The early Cold War, before nuclear parity came along in the late 1960s to turn it into a flat-out arms race, was perceived in Washington to be a battle for the hearts and minds of Europe and the rest of the free world— a battle to prevent a hostile encirclement by the Soviet Union. In those days, when Washington looked through the eyes of the world at the U.S.

atomic-energy enterprise, it saw a belligerent engine devoted to cranking out hundreds of atomic
bombs. The British were well along toward developing a power reactor fueled with natural
uranium, the kind of reactor that nations without facilities for enriching uranium might want
to buy (and which could also breed plutonium
for a British atomic bomb). The Soviet Union
had also announced a power reactor program.
"If we are outdistanced by Russia in this race,"
worried influential Senator John O. Pastore of
Rhode Island, "it would be catastrophic. If we
are outdistanced either by the United Kingdom
or France, there would be a tremendous economic tragedy to our commerce." Westinghouse
and General Electric were eager to compete in a
world market—particularly in Western Europe
and Japan—where nuclear power already ap-
peared to be competitive with coal and oil. But
the Atomic Energy Act barred private nuclear
power.

Congress proceeded to amend the old act. The
new Atomic Energy Act of 1954 allowed private
industry to own and operate reactors, though it
stopped short of allowing the AEC to subsidize
private projects beyond research and development. It encouraged private marketing activities
abroad. A GE executive summed up its effect

simply: by its provisions, he wrote, "the Government monopoly created by the 1946 Act was substantially broken."

On Labor Day in 1954, President Eisenhower started up a robot bulldozer that turned the first dirt for a 60,000-kilowatt demonstration plant at Shippingport, Pennsylvania. Duquesne Light would build the new plant jointly with Westinghouse under Hyman Rickover's direction. The reactor would be a Westinghouse Large Ship Reactor, the pressurized-water design that the engineering firm had been developing for aircraft carriers. In light of the subsequent outcry against nuclear power in the United States, the reason Duquesne decided to go nuclear is ironic. Nearby Pittsburgh had instituted strict smoke abatement and was about to restrict sulfur-oxide emissions. To the Pennsylvania utility, power from the atom looked like a godsend for pollution control.

AEC chairman Lewis L. Strauss articulated the commission's vision that year in a widely noticed speech. Predicting that nuclear power could be available commercially in five to fifteen years, "depending on the vigor of the development effort," Strauss evoked an exuberant, technologically ill-informed program of "transmutation of the elements—unlimited power . . .

these and a host of other results all in fifteen short years. It is not too much to expect that our children will enjoy in their homes electrical energy too cheap to meter, will know of great periodic regional famines in the world only as matters of history, will travel effortlessly over the seas and under them and through the air with a minimum of dangers and at great speeds, and will experience a life span longer than ours. . . . This is the forecast for an age of peace." *Energy too cheap to meter* was a phrase that would come back to haunt the new industrial enterprise.

More crucial to commercializing nuclear power was an act that Congress passed in 1957 that almost completely underwrote liability for nuclear accidents. The Price-Anderson Act required utilities to buy the maximum amount of private insurance available (which turned out to be only $65 million) but indemnified them above that maximum up to a $560 million cap. Just as the intoxicating combination of deregulation and federal deposit insurance licensed excessive risk-taking in the savings and loan industry three decades later, so the Price-Anderson Act encouraged interested utilities to rank safety issues lower on their list of responsibilities than they otherwise might have done.

Nor did the federal government emphasize

safety. "The authors of the Atomic Energy Act of 1954," writes the journalist and nuclear historian Daniel Ford, "had copied their provisions for reactor licensing almost word-for-word from the Federal Communications Act of 1934, which had established procedures for the federal licensing of radio stations. They had added a vague requirement that nuclear-plant licensees must 'observe such safety standards . . . as the [Atomic Energy] Commission by rule may establish.'" The Commission, for the next few years, largely left safety up to the utilities. In those early days of commercial nuclear power, where safety was concerned, no one was minding the store.

Government encouragement alone wasn't sufficient to persuade utilities to go nuclear. Shippingport came on line in 1957. A few other demonstration reactors followed in other places. But utilities weren't buying: reactors still cost too much to make them competitive with coal-fired plants. General Electric forced the breakthrough. The company had been developing a boiling-water reactor (BWR) in parallel with Westinghouse's pressurized-water reactor (the water circulating in a BWR serves both to moderate and cool the reactor and to make steam to run the turbine generators that make electricity). Late in 1963 GE signed a fixed-price contract with Jersey Central Power and Light to deliver a 515-megawatt turnkey plant to a site at Oyster Creek. It was the first of twelve such turnkey units that GE and Westinghouse would sell in the next three years and on which they would collectively lose nearly $1 billion. "Only extreme confidence in future technological progress," writes James Jasper, "could allow this [loss] to happen." He goes on:

The origins of this technological enthusiasm are not hard to find. . . . In 1963 both [GE and Westinghouse] got new chief officers with no technological experience: a former light-bulb salesman at GE and a financial analyst at Westinghouse. . . . With no realism to temper their technological enthusiasm, both men had unlimited confidence in the rapid development of nuclear technology. Their faith was reinforced by the rivalry between the two companies, each anxious to outpace the other. . . .

Technological enthusiasm lined up with corporate interests, and both were reinforced by the desires of two corporate heads to establish themselves in new positions. This competition between two active reactor producers is the major reason nuclear reactors were commercialized so early in the United States.

Through 1959, the AEC had spent $586 million to promote nuclear power; industry had spent only $82 million. After 1963, with GE and Westinghouse offering loss-leader reactors that they hoped declining costs would eventually make profitable, what utilities executive Philip C. Sporn christened a bandwagon market ensued. The "bandwagon effect," the skeptical president of American Electric Power Company wrote the JCAE in 1967, resulted from "many utilities rushing ahead to order nuclear power plants, often on the basis of only nebulous analysis and frequently because of a desire to get started in the nuclear business." Between 1965 and 1970, U.S. utility companies placed orders for some 100 reactors. The bandwagon rolled.

C H A P T E R 3

THE END OF
THE BINGE

Since 1978, not one U.S. utility has placed an order for a nuclear power plant, though a few
plants previously ordered but halted in midcon-
struction are currently being finished and
brought on line. Everyone who has studied the
rise and fall of nuclear power in the United
States has reached the same broad conclusion:
nuclear power stumbled to its present impasse
because the technology was commercialized too
narrowly and too soon. How manufacturers,
government regulators, and utilities all contrived
to take such an expensive wrong turn in lockstep is a classic tale of enthusiasm gone awry.
Utilities management, as we have seen, is usually

awarded the lion's share of the blame, and mismanagement was certainly a significant, perhaps
even determining, factor. But in fact, the three
groups interacted to produce the disaster.

Manufacturers were driven, partly by competitive zeal, partly by the accident of the Navy's
choice of a pressurized-water reactor design, to
develop a type of power reactor—the light-water
reactor—that was clearly less than ideal. Compared with sodium, for example, water is a less
efficient, but familiar, coolant. Uranium oxide,
standard light-water reactor fuel, is less dense
than uranium metal and conducts heat much less
efficiently. But uranium metal swells under neutron bombardment, and research scientists had
not yet had time to figure out how to compensate for the swelling problem in fuel-rod design.
Since the light-water reactor isn't a breeder, it
wastes most of its fuel, converting only about 3
percent to usable energy; that in turn increases
the volume of high-level radioactive waste. Fuel
efficiency wasn't a problem for Rickover's submarines—which, in any case, didn't have room
for the added equipment a breeder requires—
and breeding was an opportunity the manufacturers bypassed in the rush to commercialize,
because developing the more complicated technology would have taken time. The Atomic

Energy Commission compensated by selling enriched uranium for reactor fuel cheap.

To make their compromise reactor designs competitive in a field dominated by relatively cheap fossil fuels, reactor manufacturers pushed design limits. "At that time," physicist Bernard Cohen notes, "it was deemed necessary to achieve maximum efficiency and minimum cost in order to compete successfully with coal- or oil-burning plants. The latter were priced at 15 percent of their present cost and used fuel that was very cheap by current standards. In order to maximize efficiencies in the nuclear plants, temperatures, pressures, and power densities were pushed up to their highest practical limits." Tighter design limits led to more frequent shutdowns (by 1980, annual unplanned shutdowns averaged 7.4 per plant) and increased the risk of breakdown, which in turn required more complex safety systems.

More crucially, manufacturers began pursuing economies of scale vertically as well as horizontally by selling larger and larger reactors, without fully addressing the changing cost and safety issues such reactors raised. "Reactors ordered in the late 1960s were huge by historical standards, as much as seven times larger than any nuclear plant then in operation," write the policy analysts Joseph Morone and Edward Woodhouse. "The largest commercial facility operating in 1963 had a capacity of 200 megawatts; only four years later, utilities were ordering reactors of 1,200 megawatts. The average capacity of all reactors ordered rose steadily, from 636 megawatts in 1963, to 821 in 1966, to 1,029 in 1969." The safety equipment that government regulators judged sufficient at 200 megawatts they no longer judged sufficient at 1,000 megawatts; the worst possible disaster they projected, a radiation release that might induce fatal illness in 3,400 people at the smaller scale, might induce fatal illness in 45,000 people at the larger and cause as much as forty times more property damage. In consequence, government regulators began requiring further add-on safety systems as the bandwagon decade advanced, escalating engineering and construction costs. A significant portion of the increased cost of plants was additional professional labor— more engineering time—and more materials. Professional labor represented 38 percent of total labor cost in 1978; by 1987 it represented 52 percent. The big units required half again as much design work and about 40 percent more materials. "The increase in total construction

time . . . from seven years in 1971 to twelve

years in 1980," writes Cohen, "roughly doubled the final cost of plants."

Nor did the rolling bandwagon allow utilities to gain experience with operations or costs. "At no time from 1962 to 1972," James Jasper comments, "were there any plants in operation as large as the smallest of those being ordered. . . . Nuclear Regulatory Commissioner Peter Bradford . . . later commented that 'an entire generation of large plants was designed and built with no relevant operating experience, almost as if the airline industry had gone from Piper Cubs to jumbo jets in about fifteen years.' As a result . . . cost overruns beyond original estimates were greater for plants built in the 1970s than for plants built in the 1960s."

The AEC's dual role as promoter and regulator added to the confusion in the early years of nuclear power. In his promotional guise, '50s AEC chairman Lewis L. Strauss would talk of his "faith in the atomic future" and evoke a "Higher Intelligence [which] decided man was ready," to explain why nuclear fission had been discovered. Strauss, a protégé of Herbert Hoover and a former investment-banking partner at Kuhn, Loeb, believed fervently in private versus public power; he rushed private nuclear power development to steal a march on such public institutions as the Tennessee Valley Authority. Under Strauss, economics took precedence over measured development and optimal design. "I was involved very much in initiating the pressurized-water reactor," physicist Alvin Weinberg told nuclear historian Daniel Ford. "Safety was always a consideration, but I guess one would have to say that in the original conception of the light-water reactor, the primary concern—of course, it was a military device—was, Would it work? . . . There was a fundamental tug of war between *as safe as possible* and *as cheap as possible*. . . . When the nuclear business got started, the price of electricity from coal was 3 or 4 mills [tenths of a cent] per kilowatt-hour. Nuclear had to get into that ballpark, or there wouldn't be any commercial nuclear power."

Since U.S. light-water reactors aren't inherently safe designs (under some conditions of water temperature and pressure, they become more reactive), safety had to be imposed on them from the outside—"defense in depth," the AEC called it. In the late 1940s, the influential Hungarian émigré physicist Edward Teller, one of the pioneers of nuclear energy development, headed a reactor-safety committee for the AEC that recommended siting power reactors in remote areas

to protect the public from both known and un-
known risks. Keeping units relatively small also
reduced risk. Remote siting gave way to containment—confining the reactor within a thick steel
or reinforced concrete shell—when GE proposed
to build a reactor laboratory near its offices in
Schenectady, New York. To further early commercialization, the AEC endorsed the GE proposal to make use of its engineering work force
there. Utilities also turned to containment to
save transmission costs by siting reactors closer
to the cities they served. But when manufacturers
began producing larger units in pursuit of
economies of scale, the increased inventory of
hot, radioactive fuel in the larger cores raised serious concerns about potential meltdown. Under
worst-case circumstances, engineers could no
longer guarantee that containment would work.

That realization marked a major turning
point in the tragic history of first-generation nuclear power. In the mid-1960s, in response to
concerns about the safety of the new large reactors, the AEC shifted from a design philosophy
of "defense in depth," which relied on physical
containment, to one that emphasized prevention—
"engineered safety." Engineered safety meant re-
dundant operating and emergency safety systems
in addition to containment. But there was no

way to prove conclusively that such systems would always, and in all contingencies, work. Morone and Woodhouse sum up the disastrous consequences:

> The scale-up in reactor sizes forced the AEC to change its reactor-safety strategy. . . . Because containment could not be relied on to safeguard the public from especially severe accidents in large reactors, the AEC set out to prevent all such accidents. The primary approach was to upgrade emergency systems, add additional redundancies, and widen the margins for error. These changes probably did reduce the likelihood and severity of accidents. Paradoxically, however, the shift in strategy had a crucial and unexpected effect: it made it impossible to demonstrate that reactors were acceptably safe.

Utilities in turn contributed to the ongoing debacle by failing at the outset to recognize the exceptional requirements of an exotic new technology and failing to manage effectively when costs began to rise. "It's just another way to boil water," said utilities executives, minimizing the challenge of reactor operation. "The average utility knew as much about the nuclear plant it was buying as the average car-buyer knows about cars," a utility vice-president told Daniel Ford privately. "They knew how big it was, and what it cost. We got into nuclear power because the president of my utility used to play golf with the president of another utility. They bought one, and so we bought one."

The Navy's original reactor designs had emphasized efficiency over safety. To ensure safe operation, Hyman Rickover insisted upon highly trained crews. Most U.S. nuclear-plant operators moved over to nuclear work from coal and oil with minimal training, particularly training for emergencies. They were commonly only high-school graduates. "Designed by geniuses and operated by idiots" was the unfairly cruel characterization some people applied to the new industry, an image reinforced in recent years by the television cartoon character Homer Simpson, a notoriously incompetent nuclear power plant worker. Operators today are usually recruited from the enlisted ranks of men who operated nuclear submarines, but the lesson has still not been fully learned. "Nuclear power requires a dedicated organization whose only focus is the operation of those plants," Andrew C. Kadak, president of Massachusetts' Yankee Atomic Electric, remarked recently at a conference. He went on:

This is not common yet in this country, although many utilities are moving in that direction. I think the technology is important enough to dedicate essentially single-focus organizations to its operation. When the Yankee plant was built thirty years ago, the utilities who wanted to participate in the process decided they did not know very much about it, and they wanted to create a separate organization to design, build, and operate this plant. That scheme works very well.

Since Three Mile Island, U.S. nuclear power-plant operators have trained on simulators for accidents as well as for routine operation, but the Yankee Atomic Electric practice of dedicated organization was, and is, the exception among U.S. utilities.

Utility managements failed as well at cost-benefit analysis. Nuclear power was commercialized too rapidly to extract solid numbers from previous experience. According to one economic post-mortem, "a circular flow of mutually reinforcing assertions . . . apparently intoxicated both [the AEC and utility managements] and inhibited normal commercial skepticism about advertisements which purported to be analyses. . . . Even though more than 100 light-water reactors were under construction or in operation in the United States by the end of 1975, their capital cost was almost anyone's guess." Under these foggy circumstances, the finessing of insurance risk with the Price-Anderson Act was the equivalent of brake failure on a tractor-trailer rig barreling downhill. "One of the main groups in the American economic system with a persistent concern for accurate cost analyses—indeed, whose existence and success depended on them—was the insurance industry," James Jasper comments. "But it had been pushed out of the picture when the government assumed responsibility for accident liability."

The utilities industry came to these difficult new challenges from a happy land where growth flowed like a mountain river, averaging 6 to 7 percent annually since the 1950s. Costs just as reliably declined, straitened by technological improvements, so that public utility commissions were accustomed to ordering not rate increases but rate cuts. In fact, investor-owned utilities had so few restraints on growth that, in 1965, they were typically able to finance 65 percent of their capital improvements out of profits. These technological benevolences allowed industry executives to indulge a missionary vision: they believed themselves empowered to produce electricity that would continue driving the surge of growth the United States had seen since the end of World War II. When manufacturers offered reactors on terms that mollified cost analysts, management judged nuclear power to be yet another technological improvement in the ongoing story of electricity's spectacular success. Electric utilities did better in the 1960s than at any other time in their entire one-hundred-year history.

The Arab oil embargo followed in late 1973, a profound shock to the energy business, changing fundamental assumptions. In three months, oil prices soared from $3 per barrel to $11.65. Between 1973 and 1980, they increased by 1,000 percent. American industry and consumers responded by conserving across the board—electricity as well as oil—reducing overall energy demand. In 1974, there was no increase in demand, and only a 2 percent increase in 1975. Nor is demand expected to increase by more than 1.8 percent annually until the turn of the century. Energy consultant Daniel Yergin evokes the gloomy national mood the oil shock provoked:

> The shortfall struck at fundamental beliefs in the endless abundance of resources, convictions so deeply rooted in the American character and experience that a large part of the public did not even

know, up until October 1973, that the United States imported any oil at all. . . . The embargo and the shortage it caused were an abrupt break with America's past, and the experience would severely undermine Americans' confidence in the future.

Faced with an abrupt drop in demand and rapidly escalating construction costs, hard-nosed accounting reasserted itself in the nuclear power industry. The surge of building had made the utilities dependent on the financial markets for development capital—financing from income dropped from a 65 percent share in 1965 to less than 30 percent in the 1970s—and investment was no longer forthcoming. A Brookings Institution study found that by the end of 1974, utilities had canceled or deferred 235 nuclear and fossil-fuel plants, a total of nearly 200,000 megawatts. "Complex econometric analyses," Jasper notes, "well designed to capture changes in demand, began to replace the extrapolations of past trends that fit so well with the technological enthusiasm characteristic of electric utilities." The binge was over; the utilities sobered up.

C H A P T E R 4

OTHER VOICES, OTHER ROOMS

Ambassador Ryukichi Imai, doctor of engineering and senior adviser to the Japan Atomic Industrial Forum, is a small, solid Japanese diplomat with a round face and a nonchalant Harvard manner. At lunch in an elegant restaurant overlooking a park near the center of Tokyo, he orders a Suntory on the rocks and remarks that he no longer frequents the park at night—his colleagues on the executive floors of the surrounding office buildings, he observes deadpan, scan it with night-vision scopes, looking for lovers, the latest Japanese high-tech fad.

The ambassador—to Kuwait, to international arms-control and disarmament conferences, to

Mexico—remembers the early days of Japanese nuclear power. Interest surfaced at the Atoms for Peace conference that the Eisenhower administration sponsored in 1955, he says. Many nations came to that historic gathering, the first open postwar conference devoted to nuclear energy, and everyone had something to put on the table: a paper, some sort of report, some information, some research. Japan had nothing and was extremely embarrassed. Imai's country was just at the beginning of its great economic takeoff, having become, he quips, "this supply warehouse" for the United States during the Korean War. When the Japanese delegates got home from the conference, they discussed what they should do. "And Nakasone," growls Imai in his unaccented English, referring to the Liberal Democrat who would later become a notable prime minister of Japan, "Nakasone said, 'For a change let's not form a commission; let's get started.'" Japan's Science and Technology Agency (STA) got its first nuclear budget that same year, ¥235 million. Imai swears they pulled the number out of a hat. "It's the isotope number for uranium-235," he chuckles.

Today the STA budget for nuclear power is more than ¥400 billion ($3.2 billion), and 41 commercial power reactors in Japan supply

33,239 megawatts of electricity, some 27 percent of the nation's total electrical supply, making Japan the fourth-largest generator of nuclear power in the world (after the United States, the former Soviet Union, and France). Nuclear power has become Japan's lowest-cost source of electricity, and the Japanese estimate that it will become their cheapest energy source overall by 2010, when they hope to have built 40 more units to boost the nuclear share of national electrical production to 43 percent.

If Japan took up atomic energy in the early days for reasons of prestige, as Ambassador Imai implies, it realized after the Arab oil embargo that nuclear power was the only reliable answer to its energy needs. The small island nation is fourth in the world in per capita energy consumption; because it lacks fossil fuel resources, 80 percent of that energy is imported, predominantly in the forms of Middle Eastern oil and natural gas and Chinese coal. Since 1973 the nine Japanese utility companies have responded to increasing consumer and industrial electrical demand by building an average of more than two nuclear power plants per year. The Japanese purchased their first commercial reactor—a gas-cooled, graphite-moderated unit, fueled with natural uranium—from the United Kingdom but didn't like the technology. As soon as Westinghouse and General Electric began offering pressurized-water and boiling-water reactors for export, the Japanese bought them, and they've been building them in Japan under license ever since. Along the way they've made their industry the safest in the world, with no major accidents and fewer unplanned shutdowns per reactor than any other country, 0.4 per year, compared with 4.0 in the United States. In 1991 I went to Japan to find out why the Japanese coax better performance from American machines than Americans do.

On a cool day in early spring—the plum blossoms were just losing their bloom on the grounds of Kyoto temples—I rode a Shinkansen bullet train an hour out of Kyoto to Fukui Prefecture on the west coast of Honshu; from there a car took me on to Japan Atomic Power Company's Tsuruga Power Station on rugged Urasoko Bay. Tsuruga Unit 1, the first commercial light-water reactor in Japan, began operation in 1970. I had come to see the newer unit set into the hill saddle behind it, Tsuruga Unit 2. The 1,160-megawatt PWR went on line in 1987, the first model plant in an industry-wide program of improved and standardized design incorporating Japanese technology. Japan is an archipelago of narrow, mountainous islands, short on flat land; once a Japanese utility has gone through the expensive, time-consuming process of acquiring a reactor site, it builds as many reactors on the site as possible. Along the same stretch of coast, tucked away in bays, no fewer than 15 commercial power reactors are operating or under construction, including Monju, Japan's first fast-breeder reactor. The remote coast, over the mountains beyond Kyoto, is safely distant from Japan's major population centers—three hours by Shinkansen from Tokyo, on the other side of Honshu.

Yasuo Sasaki, Tsuruga's executive manager and chief engineer, met me in the lobby of the power station's sparkling, modern public relations pavilion. Tall for a Japanese of his generation, with a broad, flat face and no-nonsense glasses, Sasaki was wearing the same basic light-gray, yellow-trimmed uniform that I would later see on his operator crews. Management and staff dress alike at Tsuruga, change in the same locker rooms, and eat in a common dining hall, and it was clear along the way that Sasaki is highly respected: he earned a full complement of admiring bows wherever he passed.

The PR pavilion included a small science museum, part of a continuing program of education designed to inform the Japanese people about nuclear power. One exhibit made a point-by-point comparison between the Chernobyl RBMK-type reactor that exploded in 1986 and a Japanese PWR to demonstrate the PWR's greater inherent safety. A joy stick at another exhibit allowed visitors, including thousands of schoolchildren every year, to direct a remote color-video camera sited on a nearby pond; one of Sasaki's aides quickly brought up a brace of nesting cranes on the monitor. The exhibit quietly asserts that nuclear power can be compatible with its natural surroundings. Tsuruga even maintains a bird-watching cabin near the pond for local organizations to use. Antinuclear activists think such exhibits propaganda, but there was nothing false in either one. Environmental monitoring around Tsuruga is extensive, and even the warm seawater effluent from the turbine-cooling heat exchangers supports an experimental fish farm farther down the bay.

Similar education is ongoing throughout the country. Toichi Sakata, director of the nuclear-fuel division of the Science and Technology Agency, told me a few days later in Tokyo that STA has recently provided about 100 nuclear scientists to speak to local communities and local clubs such as Rotary—to grass-roots groups of all kinds. "So far we have sent these nuclear experts 260 times to speak to about 13,000 people," Sakata said. "We're also organizing seminars for local leaders." Tours of nuclear facilities, Sakata suggested, were "the most accepted way to educate people to what is nuclear power."

Acquiring a site for a nuclear power plant in Japan can take as long as twenty years. As in so many other aspects of Japanese life, utilities work toward local consensus. An American knowledgeable about the Japanese nuclear power industry told me that there's much wining and dining of local politicians and money changing hands under the table. "Fishermen insist on compensation for our discharging warm water into their fishing grounds," a Japanese utility executive commented, even though fish grow faster in warmer water. Fishermen get compensation; every nearby village gets a new public hall. "Once a local community accepts nuclear facilities," the STA's Sakata explained, "they are entitled to receive special financial support from the government. They can use that money in their local communities to build roads, to build public halls. Without getting positive support from local governments, nuclear industries can't build." Once support is forthcoming, the spread between start of construction and start of commercial operation averages only four or five years, less than half the time construction has required in the United States.

Executive manager Sasaki outfitted me with a standard yellow hardhat and took me on a tour of Tsuruga Unit 2. Going in, I failed the security-gate test, to his great embarrassment. The double-doored steel passage, controlled by a pass card on the outside and a security code within, also featured a built-in alarm triggered by weight to guard against the possibility of two people slipping through on one card and code, but the scale was set to Japanese averages. I'm a sturdy 190-pound American, and I made all the alarm bells ring. A fire team in silver suits went roaring by at the same time on a fire drill that was probably staged for my benefit but impressive nonetheless.

The reactor building was well-organized and spotless, as I expected it to be. I remarked to Sasaki that American power plants weren't so clean, and he shuddered agreement—he'd seen them. He kept his plants clean, he said, not only to keep down radioactivity but also to build morale. Everyone who works at Tsuruga receives a monthly printout of his radiation exposure; the lower the dosage, the better employees are likely to feel about their work. Average worker exposure to radioactivity has declined every year since 1979 in Japan, even as the number of commercial reactors has nearly doubled.

Later during my visit, a contact would pass me a copy of a confidential report prepared by an American investigator in 1984 for the U.S. Nuclear Regulatory Commission. The report sought to discover why Japanese nuclear power plants had better operating records than their U.S. counterparts. Among its conclusions were these:

- Japanese requirements call for a basic maintenance plan to be submitted and approved by the regulatory body. No corresponding requirement exists in the United States, nor is there a requirement in the U.S. for NRC personnel to witness the actual inspections . . . as inspectors do . . . in Japan.
- The housekeeping in the plants visited was outstanding.
- The Japanese approach . . . concentrates on reducing the maintenance workload through automation and increases the ease of maintenance by providing greater accessibility to equipment rather than through the application of foolproof design.

- Coding and labeling of equipment within Japanese plants is excellent and far exceeds that typically found in plants in the United States. This includes the use of location aids throughout the plant, color banding on piping to indicate the fluid transported by the piping and the system [to] which the piping belongs. Labeling is used extensively. . . .

- Most (70 percent) of the maintenance during the annual inspection is carried out by contractor personnel who specialize in such maintenance.

Color coding, labeling, and accessible equipment are simple and homely remedies, but the difference between their success and the arrogance of the phrase *foolproof design* speaks volumes.

The Tsuruga Unit 2 reactor is solidly encased in the first prestressed-concrete containment vessel in Japan, improving seismic stability in that earthquake-prone country (earlier vessels were welded steel). Building layout was reorganized to reduce the number of structures from the typical six or seven to only three, another innovation that improves earthquake resistance. The control room, as I saw, is greatly simplified and improved over typical U.S. designs—clearly the Japanese also learned from Three Mile Island.

Other significant innovations at Tsuruga Unit 2 aren't visible on a tour. The Japanese have added waste-treatment systems to their commercial reactors to reduce the release of radioactivity into the environment even further. Gases go into hold-up tanks to allow them to decay to safe levels before they're released. Liquids go through charcoal, electromagnetic, or ion-exchange filtration before discharge or reuse. Other liquids are condensed by evaporation and then solidified with asphalt and barreled for storage underground.

None of these differences between Japanese and U.S. operation quite explains the large gap in operating efficiency and safety between the two countries. It helps that there are only nine utility companies in Japan compared with the legion in the United States; the Japanese utilities benefit from greater continuity of experience and have the financial depth to support long-term development that includes a primary emphasis on safety. It helps that the government closely monitors, hands-on, their nuclear operations. The Japanese, like the French, benefited from starting late in the nuclear power business, leaving the inevitable mistakes of pioneering to the United States. Pioneering should have been an advantage to the latter, but poor management frittered the advantage away.

Japanese utilities aren't immune to the kind of quality-control problems that plagued U.S. utilities when they were still building reactors. In February 1991, a small steam tube burst in one of the three reactors at Mihama, north of the Tsuruga power station on the west coast of Honshu, and the Japanese experienced their first unplanned emergency core-cooling-system activation. The Mihama reactor will be down for several years while its steam generator is replaced, as will a matching reactor at Takahama (in central Japan near Kyoto), subsequently identified as having the same problem. A safety inquiry turned up the surprising information that the Mihama tube burst because the antivibration brackets to which it and many other small tubes were supposed to be attached had been badly installed. Twenty years earlier, workmen had apparently had difficulty fitting the brackets into the allotted space and vented their frustration by leaving the tubes unsupported, sawing off the protruding ends of the brackets, and jamming the cover panel shut. That left the stainless steel tubes vibrating in the high-velocity flow of cooling water around them, resulting eventually in metal fatigue. What amounted to industrial sabotage resulted in a potentially dangerous incident twenty years later. A spokesman for Japan's Ministry of International Trade and Industry explained after the accident that the damage had escaped detection for all those years because "no one ever imagined the antivibration bars were incorrectly installed."

The Japanese feel greater pressure to make nuclear power work given their poverty of fossil fuels—it's meat and potatoes for them, not icing on the cake. The pressure's increasing. The Japanese people are trading in their residential ceiling fans for air-conditioning units in record numbers. The hot summer of 1990 saw a near blackout in Tokyo. At our luncheon, Ambassador Imai leaned across the table in dead earnest to tell me about it. "August," he said, "the height of the hot summer season, everyone had his air conditioner on, everyone was home watching the national high-school baseball championships, and then at five o'clock, everyone plugged in his rice cooker." He paused dramatically. "It was a very near thing."

Though they are building commercial nuclear reactors as fast as they can and installing gas-turbine systems to help with peak loads, the Japanese are faced with a continuing crisis of electrical supply and demand, despite a strenu-

ous program of conservation that proposes converting half the residential water heaters in the entire nation to solar by 2010. Even finding sites for the many new nuclear units they estimate they need is problematic on their crowded islands.

"The Japanese are the Swiss of Asia," a U.S. businessman finding wisdom at the bottom of a sake cup told me one evening in Tokyo. The difference between U.S. and Japanese commercial nuclear power operation isn't mysterious after all. The Japanese are careful and conservative in their engineering, learn from experience, resist technological arrogance, and use highly trained crews. When an engineered system isn't inherently safe—and no system as complex as a nuclear reactor can ever be completely safe, completely "foolproof"—then knowledge, training, high morale, and watchful respect must substitute for physical safeguards. Hyman Rickover knew that; so do the Japanese.

The French, equally impoverished of fossil-fuel resources, had more trouble getting started in the nuclear power business than the Japanese. The trouble was primarily political—conflicts within the national bureaucracy and resistance at some proposed construction sites. Once those difficulties were resolved, in the late

1970s, nuclear power in France moved forward rapidly and now accounts for 75 percent of French electrical output, making France the most highly nuclear-powered country in the world, with 54 reactors generating 63,000 megawatts of power. The United States doubles both those numbers, but in terms of per capita consumption the difference is dramatic: per French citizen 4,931 kilowatt-hours, per U.S. citizen 2,265 kilowatt-hours, per Japanese citizen 1,450 kilowatt-hours. In France the slogan has long been "tout électrique, tout nucléaire"—"all electric, all nuclear." With nuclear power and stringent conservation, the French have been able to reduce their dependence on imported energy sources from 77 to 50 percent (and their air pollution by a factor of five). At the same time, putting all their eggs in a nuclear basket has become worrisome in the wake of Chernobyl. "A new Chernobyl would probably be the end of nuclear energy in Europe," Pierre Tanguy, the inspector general for nuclear safety at Electricité de France (EDF), France's national public utility, told me when I interviewed him in his Paris offices.

"In France in the years immediately after the war," said Tanguy, "it was quite clear that we were cheated by the British about the oil in the

Mideast. We had no oil. There was this feeling very deep in the mind of the French people that dependence on foreign energy could be a source of trouble, not only economic trouble but political trouble as well. So immediately after the war, and it was right from 1945, there was the establishment of CEA, the atomic energy commission."

CEA wanted more than an independent nuclear power industry, however. "Its first objective," a French government official admitted, "was to have the atom bomb. The second objective was to develop a nuclear reactor. So it developed a French nuclear power reactor, gasgraphite." The gas-graphite reactor used natural uranium—France had not yet developed facilities for uranium enrichment—and one of its important by-products was plutonium for bombs. Gabon and South Africa supplied the necessary ore.

In 1960 in the Sahara, France exploded its first atomic bomb, a 60-kiloton plutonium implosion weapon. By then one pressurized-water and three gas-graphite commercial reactors were under construction around the country. A total of six gas-graphite reactors went on line before the 1973 Arab oil embargo. In a classic battle over bureaucratic turf, EDF was nurturing France's nuclear power ambitions during those early years, CEA resisting them. "The French government does not have an energy policy," EDF director general Marcel Boiteux is said to have told close associates in 1970. "I am obliged to have one in its place."

Boiteux is an economist, a cool and forceful man; he headed EDF from 1967 to 1979. I interviewed him in Paris. When France decided to go nuclear in 1974, he explained, the decision was based on economics. "Three reasons justified the 1974 program," he said: "Dependence on oil, the high cost of national coal production, and the profitability of nuclear production."

Tanguy sums up France's choices this way: "Lord Marshall, who used to be the chairman of the British counterpart to EDF, said there were four reasons why France had to choose the nuclear option: France has no coal, France has no gas, France has no oil, France has no choice." But it's also true that EDF saw a chance to expand its authority within the French government and stacked the deck of a crucial government committee, persuading or appointing a majority of the committee's membership. The Messmer Plan of 1974, named after French prime minister Pierre Messmer, called for construction to begin on thirteen new reactors before the end of 1975.

They would be Westinghouse pressurized-water

reactors. The French were willing to make the switch to a system that used enriched uranium, Boiteux told me, because their domestic uranium-enrichment plant was coming into service. "It would have been ridiculous to try to escape from oil dependence just to create a dependency on foreign enriched uranium."

Going nuclear wasn't easy at first. "There were enormous problems between 1974, when the program was launched, and 1977," says Boiteux:

Our employees received death threats. Coffins were delivered to the plant sites. My apartment was bombed with *plastique*—the stairs collapsed through eight floors. It was a very difficult time. At the end of July 1977, the president of the republic, Giscard d'Estaing, courageously—because he was the first person to do so—announced that the nuclear policy was not an EDF policy, it was a French policy. And that changed the climate completely, because once the whole of the political scene had taken a positive position in relation to nuclear power, there was little protest.

In the thirteen years from 1974 to 1987, France ordered, installed, and began operating twenty-nine 900-megawatt PWRs, twenty 1,300-megawatt PWRs, and two 1,400-megawatt PWRs, each line built of standardized subunits. The 1,400-megawatt series is an indigenous French product with more efficient turbines and totally computerized control rooms. Between 1987 and 1991, orders ceased because the industry found itself with seven or eight plants' worth of surplus capacity. With increasing demand, it expects to absorb that capacity by 1995. EDF ordered one new unit in 1991 and will continue to order one a year until 2000. After the turn of the century, because it will have to begin decommissioning its oldest 900-megawatt units, it will build two new units per year. In the meantime it has been converting its 900-megawatt units to operate on mixed-oxide (MOX) fuels of 5 percent plutonium oxide and 95 percent depleted uranium oxide, allowing EDF to begin using the extensive stockpile of plutonium and depleted uranium (depleted of U-235) it has accumulated from fuel reprocessing.

Standardization bought the French significant savings; their reactors cost an average 25 percent less to build than those of any other country. Critics have charged that standardization is dangerous. "They say that if we have a safety problem at one plant," comments EDF's Tanguy, "that means that potentially we will have the same problem in all other units. They say that this is the dark side of standardization. My personal opinion is exactly the opposite. I think it's also the bright side. Because when we have a problem in one unit, we know where to look in the other units. All the units aren't of the same age. So we know what is going to happen to us down the road if we don't make repairs. I think that's an advantage." Standardization also shortens construction schedules and reduces costs, improves personnel training and helps ensure equipment reliability. The French shutdown record is not as good as the Japanese—the 900-megawatt units averaged 1.8 unplanned shutdowns in one recent year, the 1,300-megawatt units 4.3—but France's overall safety record is as good as any in the world.

Although I went to France to investigate the entire French nuclear program, I was particularly interested in what is called the "back end" of the nuclear-fuel cycle, France's successful program of reprocessing fuel and of confining and storing waste. In earlier years at Marcoule, in the south, and since 1970 on a much larger scale at La Hague, overlooking the English Channel near Cherbourg, the CEA-owned industrial group Cogema has reprocessed spent-fuel assemblies to separate reusable uranium and plutonium from fission by-products and other waste. Eighty percent of the oxide reactor fuels reprocessed worldwide in the past ten years were reprocessed at La Hague, some 850 tons per year.

I flew from Paris to Nîmes and drove through wine country to Marcoule, near Avignon, to visit the waste processing and storage facility there. Marcoule has processed all the fuel waste from France's six gas-graphite reactors; the technology developed there has been expanded into a major operation at La Hague and exported to Japan for the reprocessing and storage facility the French built at Tokai-mura, which went on line in 1977.

Using uranium for reactor fuel on a one-time-through basis, as the U.S. nuclear power industry does, is extremely inefficient. Only about 3 percent of the U-235 burns up under such a regimen, and the plutonium bred from U-238—the bulk of the fuel—goes to waste. Hundreds of tons of valuable "spent" fuel elements sit in storage pools at commercial reactor sites throughout the United States. Reprocessing could salvage all but a small part of those fuel elements and dramatically reduce the volume of material that would need to be buried as waste. But reprocessing has a bad name in the United States. Like nuclear power itself, it was commercialized prematurely here; in its pioneer form it was both expensive and polluting, and the last commercial reprocessing plant built in the United States was closed down twenty years ago.

The French reprocess for themselves as well as for Japan, Germany, Switzerland, Belgium, and the Netherlands, separating their used-fuel elements into plutonium, depleted uranium, fission products, and irradiated structural materials. In the French case, the plutonium and depleted uranium go back together to form MOX fuels which are then reused. Structural materials are encased in concrete for underground storage. What's left is a liquid residue of highly radioactive fission products. These products are the most biologically dangerous materials a reactor produces, but their radioactivity is relatively short-lived. Burying unprocessed spent fuel, as the United States proposes to do, means burying plutonium and uranium, elements with half-lives measured in tens of thousands and millions of years. Because these time-spans far exceed recorded human experience, critics charge that no burial plan can guarantee such waste will remain isolated from the environment—at least not long enough to decay to levels equivalent to the natural background of radiation always present in the earth's crust and therefore presumed safe. By contrast, separated fission products lose

99 percent of their radioactivity in about six hundred years—still a long time-span, but one well within the scale of human history. The Pantheon in Rome and many other structures have been preserved for that length of time and longer on the surface of the earth; rocks buried deep underground typically last tens of millions of years, even without the special barriers between them and the environment such as buried waste will have.

At Marcoule I watched through the thick yellow windows of a hot cell as a rotating steel cylinder, called a calciner, reduced concentrated liquid fission products to a fine powder. The powder was then mixed in an electric furnace with glass frit to form a red-hot glass. The viscous glass poured in a slow, glowing stream into a stainless-steel tank like a scaled-up version of an old-fashioned milk can. When the glass cooled it would be black as obsidian and radioactive; a cylinder of such material not much larger than a shot glass would incorporate all the fission products that resulted from producing a lifetime supply of electricity for a French family of four.

The process I watched is called vitrification. Once the glass had cooled in its stainless-steel can, a shielded robot-transfer cask, painted bright red, would socket itself into one wall of the hot cell, swallow the can, and move it out into the adjoining storage hall. The Marcoule vitrification storage hall is a building the size of a small warehouse. Its concrete floor is patterned with red and green manhole-size concrete plugs that seal off temporary storage shafts descending several hundred feet underground. Vitrified in the steel cans in those shafts was the entire volume of reduced waste from the first decades of French nuclear power. I stood on that polka-dotted floor and listened to the slow clicking of a Geiger counter in one corner of the hall as it recorded only the natural background of radiation. When France finally opens a permanent waste-disposal site, the Marcoule cans will be moved there and stored permanently underground. It isn't true, as critics claim, that we don't know how to store nuclear waste. The technology is available; the problem is trivial. What is true is that storing unprocessed so-called spent fuel, as the United States proposes to do, is a prodigal waste of valuable nuclear fuel and a long-term maintenance problem that the nation doesn't need.

Near the end of our interview, EDF's Pierre Tanguy addressed the challenge of nuclear management that has progressed so differently in France and the United States. "The phrase *safety culture* is used in this field," he told me, "but I think it's first of all *nuclear culture*." He went on:

> *Nuclear* is different. I've been working in the nuclear business now for thirty-seven years—which is quite a lot—and from time to time, you find people trying to demonstrate that, after all, the nuclear industry is just another industry. It's not true, not only because the public perception is different but also because the technical approach is different. Say we have a problem in a nuclear area. It would be trivial in a conventional fossil-fuel plant. You would just send for a good mechanic. When the problem is nuclear, you need a special organization, a special laboratory, you need some robotics to take a sample, you need some special nondestructive control tools which are really unique—all that, and support from an engineering organization as well. In a nuclear utility you need someone very close to the CEO of the company who can be in charge of the nuclear business, just so the people realize that the top manager is directly interested in the result. If you don't do that, then there will be complacency.

"Complacency in a conventional plant," Tanguy concluded, echoing Stone & Webster's

Don Wille, "means the loss of a few points of availability. In a nuclear plant, it could mean the loss of your capital, of your investment. It could mean Three Mile Island."

CHAPTER 5

A MATTER OF RISK

One minute past 4 a.m. on Wednesday, March 28, 1979, maintenance workers cleaning sludge from a small pipe blocked the flow of water in the main feedwater system of the Unit 2 reactor at Three Mile Island on the Susquehanna River near Harrisburg, Pennsylvania. The shift foreman heard "loud, thunderous noises, like a couple of freight trains," coming from Unit 2. Loudspeakers broadcast the warnings "Turbine trip . . . reactor trip" as the TMI reactor and the huge steam turbine it drove automatically shut down. Since the reactor was still producing heat, it heated the blocked cooling water around its core hot enough to create a pressure surge, which popped a relief valve. Three emergency feedwater pumps started up to restore circulation. At that point all should have been well.

But the relief valve stuck open, and some 220 gallons of water per minute began flowing out of the reactor vessel. Two valves that normally channeled water from the emergency feedwater pumps into the system were closed—probably left that way inadvertently during routine testing two days before. Other emergency pumps on the system could have supplied the reactor vessel with enough cooling water to replace the escaping water, but the control-room operators didn't know that the valve was stuck open. Four minutes after the reactor trip, believing that the reactor had too much water rather than too little, they shut these emergency pumps down. Nuclear historian Daniel Ford describes what followed:

Within five minutes after Unit 2's main feedwater system failed, the reactor, deprived of all normal and emergency sources of cooling water, and no longer able to use its enormous energy to generate electricity, gradually started to tear itself apart. The pressure of the water inside, which had increased suddenly in the few seconds after the accident began, now kept decreasing uncontrollably. The water remaining inside the reactor began to flash into steam, which in the next few hours expanded and blanketed much of the reactor's fuel rods, preventing effective cooling. A rash of further instrument problems, equipment malfunctions, computer-system breakdowns, and other difficulties beset Unit 2 operators and the growing number of plant officials who were coming to their aid. "It seemed to go on and on, surprise after surprise," Thomas Mulleavy, a radiation-protection supervisor at the plant, has recalled. "The equipment that we had to use did indeed malfunction, as most equipment will do on occasion, and always seems to when you need it most." At 7:24 a.m., shortly after Gary Miller, the station manager, arrived in the Unit 2 control room, a "general emergency" was declared—the first ever to arise at a commercial nuclear power plant in the United States.

The loss of coolant at Unit 2 continued for some 16 hours. About a third of the core melted down. Radioactive water flowed through the stuck relief valve into the containment building and then into an auxiliary building, where it pooled on the floor. Radioactive gas was released into the atmosphere. NRC officials feared that a hydrogen bubble was developing in the Unit 2 reactor vessel that might blow open the vessel itself. An estimated 140,000 people evacuated the area. President Jimmy Carter visited the scene by air, focusing the nation's attention. It took a month to stabilize the malfunctioning unit at TMI and safely shut it down. The reactor was a total loss; the cleanup required years and cost hundreds of millions of dollars.

Three Mile Island was a serious accident from the point of view of its owner, General Public Utilities. In Pierre Tanguy's phrase, it meant the loss of the utility's capital, of its investment. From the public's point of view, however, TMI was a far less threatening event than it seemed at the time. The radioactive water remained confined to the plant. The radioactive gas releases, never significant, were quickly diluted and dispersed. The hydrogen bubble was overrated; there was never enough oxygen within the reactor vessel to allow the hydrogen to ignite, and even if it had done so, post-accident analysis concluded that it would not have breached the reinforced-concrete confinement structure enclosing the reactor.

No such accident is ever desirable, but the accident at TMI had beneficial consequences for U.S. nuclear-reactor safety; it led to greater realism about potential risks, improved regulation, improved safety systems, and improved operator training and supervision. As a result of TMI, according to physicist Bernard Cohen, "the average number of operating hours per year for [U.S.] plants has increased by 12 percent. Unplanned shutdowns have been reduced by 70 percent, accident rates for workers have declined more than threefold, the volume of low-level radioactive waste has declined by 72 percent, and radiation exposure to workers has been cut in half."

TMI inspired and the NRC mandated safety modifications to nuclear plants throughout the United States that averaged $20 million per plant. "It is not an exaggeration to say," Cohen concludes, "that lessons learned from the Three Mile Island accident revolutionized the nuclear power industry."

A far more serious accident occurred seven years later at Chernobyl, in what was then still the Soviet Union. At the time of the accident—April 26, 1986—the Chernobyl nuclear power station consisted of four operating 1,000-megawatt power reactors sited along the banks of the Pripyat River, about sixty miles north of Kiev in the Ukraine, the fertile grain-producing region of the southwestern USSR. A fifth reactor was under construction.

All the Chernobyl reactors were of a design that the Russians call the RBMK—natural uranium-fueled, water-cooled, graphite-moderated—a design that American physicist and Nobel laureate Hans Bethe has called "fundamentally faulty, having a built-in instability." Because of the instability, an RBMK reactor that loses its coolant can under certain circumstances increase in reactivity and run progressively faster and hotter rather than shut itself down. Nor were the Chernobyl reactors protected by containment structures like those required for U.S. reactors, though they were shielded with heavy concrete covers.

Without question, the accident at Chernobyl was the result of a fatal combination of ignorance and complacency. "As members of a select scientific panel convened immediately after the. . . accident," writes Bethe, "my colleagues and I established that the Chernobyl disaster tells us about the deficiencies of the Soviet political and administrative system rather than about problems with nuclear power."

The immediate cause of the Chernobyl accident was a mismanaged electrical-engineering experiment. Engineers with no knowledge of reactor physics were interested to see if they could draw electricity from the turbine generator of

the Number 4 reactor unit to run water pumps
during an emergency when the turbine was no
longer being driven by the reactor but was still
spinning inertially. The engineers needed the reactor to wind up the turbine; then they planned
to idle it to 25 percent power. Unexpected electrical demand on the afternoon of April 25 delayed the experiment until eleven o'clock that
night. When the experimenters finally started,
they felt pressed to make up for lost time, so they
reduced the reactor's power level too rapidly. That
mistake caused a rapid buildup of neutron-
absorbing fission by products in the reactor
core, which poisoned the reaction. To compensate,
the operators withdrew a majority of the reactor's control rods, but even with the rods withdrawn, they were unable to increase the power
level to more than 30 megawatts, a low level of
operation at which the reactor's instability potential is at its worst and that the Chernobyl
plant's own safety rules forbade.

At that point, writes Russian nuclear engineer
Grigori Medvedev, "there were two options: increasing the power immediately, or waiting
twenty-four hours for the poisons to dissipate.
[Deputy chief engineer Dyatlov] should have
waited. . . . But he [had an experiment to con-
duct and he] was unwilling to stop. . . . He

ordered an immediate increase in the power of the reactor." Reluctantly the operators complied. By 1 a.m. on April 26, they stabilized the reactor at 200 megawatts. It was still poisoned and increasingly difficult to control. More control rods came out. A minimum reserve for an RBMK reactor is supposed to be 30 control rods. At the end, the Number 4 unit was down to only six control rods, with 205 rods withdrawn.

The experimenters allowed this dangerous condition to develop even though they had deliberately bypassed or disconnected every important safety system, including the emergency core-cooling system. They had also disconnected every backup electrical system, down to and including diesel generators, that would have allowed them to operate the reactor controls in the event of an emergency.

At 1:23 in the morning, the engineers proceeded with their experiment by shutting down the turbine generator. That reduced the electrical supply to the reactor's water pumps, which in turn reduced the flow of cooling water through the reactor. In the coolant channels within the graphite-uranium fuel core, the water began to boil.

Graphite facilitates the fission chain reaction in a graphite reactor by slowing neutrons.

Coolant water in such a reactor absorbs neutrons, thus acting as a poison. When the coolant water in the Number 4 Chernobyl unit began turning to steam, that change of phase reduced its density and made it a less effective neutron absorber. With more neutrons becoming available and few control rods inserted to absorb them, the chain reaction accelerated. The power level in the reactor began to rise.

The operators noticed the power surge and realized they needed to reduce reactivity quickly by inserting more control rods. They hit the red button of the emergency power-reduction system. Motors began driving all 205 control rods as well as the emergency protection rods into the reactor core.

But the control rods had a design flaw that now proved deadly: their tips were made of graphite. The graphite tips attached to a hollow segment one meter (3.28 feet) long, which attached in turn to a five-meter absorbent segment. When the 205 control rods began driving into the surging Number 4 reactor, the graphite tips went in first. Rather than reduce the reaction, the graphite tips increased it. The control rods displaced water from the rod channels as well, increasing reactivity further. All hell broke loose—the reactor exploded.

The explosion was chemical, driven by gases and steam generated by the core runaway, not by nuclear reactions; no commercial nuclear reactor contains a high enough concentration of U-235 or plutonium to cause a nuclear explosion. Medvedev, who had once worked at Chernobyl and who was on the scene within days, describes the explosion from the testimony of eyewitnesses:

Flames, sparks, and chunks of burning material went flying into the air above the Number 4 unit. These were red-hot pieces of nuclear fuel and graphite, some of which fell onto the roof of the turbine hall where they started fires. . . . About 50 tons of nuclear fuel evaporated and were released by the explosion into the atmosphere. . . . In addition, about 70 tons were ejected sideways from the periphery of the core, mingling with a pile of structural debris, onto the roof . . . and also onto the grounds of the plant. . . .

Some 50 tons of nuclear fuel and 800 tons of reactor graphite . . . remained in the reactor vault, where it formed a pit reminiscent of a volcanic crater. (The graphite still in the reactor burned up completely in the next few days.)

The resulting radioactive release, Medvedev estimates, was equivalent to ten Hiroshimas. In fact, since the Hiroshima bomb was an air-burst—no part of the fireball touching the ground—the Chernobyl release polluted the countryside much more than ten Hiroshimas would have done.

No commercial reactor in the United States is designed anything like the RBMK reactor. Cohen summarizes several of the differences:

1. A reactor which is unstable against a loss of water could not be licensed in the United States.
2. A reactor which is unstable against a temperature increase could not be licensed here.
3. A large power reactor without a containment [structure] could not be licensed here.

The absence of a containment structure is especially important. As Cohen points out about Chernobyl, "Post-accident analyses indicate that if there had been a U.S.-style containment, none of the radioactivity would have escaped, and there would have been no injuries or deaths."

But if the design of Russian and U.S. reactors is critically different, broad similarities between the two countries' management of nuclear-power development led both national programs into difficulty. In the U.S.S.R., writes Medvedev, "the ordinary citizen was made to believe that the peaceful atom was virtually a panacea and the ultimate in genuine safety, ecological cleanliness, and reliability." He quotes Soviet scientists and managers who waxed as enthusiastic in the heyday of nuclear power development as the U.S. AEC's Lewis Strauss. "Nuclear power stations are like stars that shine all day long!" academician M. A. Styrikovich claimed in 1980. "We shall sow them all over the land. They are perfectly safe!" The deputy head of the State Committee on the Utilization of Nuclear Energy, notes Medvedev, told the Soviet people that "nuclear reactors are regular furnaces, and the operators who run them are stokers"—an image corresponding to the glib coinage in the United States that nuclear power is "just another way to boil water."

Given such uninformed enthusiasm for technology, it isn't surprising that both the Soviet and U.S. nuclear power programs ran into difficulties, or that the difficulties in both cases were predominantly managerial. Nuclear power came to terrible disaster in the former Soviet Union because authority dominated there to the exclusion of informed technical discussion and judgment. "Accidents," writes Medvedev, "were hidden not only from the general public and the government but also from the people who worked at Soviet nuclear power stations. This latter fact posed a special danger, as failure to publicize mishaps always has unexpected consequences: it makes people careless and complacent."

Authority dominated in the early days of nuclear power in the United States as well. "The AEC and the JCAE," James Jasper notes, "placed themselves outside normal political accountability." Fortunately, both public and private sectors of the U.S. nuclear power industry learned the lessons of Three Mile Island and launched a major effort of improvement and regulation.

Three Mile Island and Chernobyl represent extreme instances of the problem that seems to trouble the American public more than any other about commercial nuclear power: its apparent danger. But risk is always relative. Friend and foe have estimated the relative risk of operating commercial nuclear power plants in the United States; their conclusions are instructive.

The most serious example of public exposure to radiation from a nuclear power plant is, of course, Chernobyl. The explosion at Chernobyl blew radioactive gas and dust high into the atmosphere, where winds dispersed it across

Finland, Sweden, and central and southern Europe. "The sum of [Chernobyl] exposures to people all over the world," writes Bernard Cohen, "will eventually, after about fifty years, reach 60 billion millirems, enough to cause about 16,000 deaths." (Millirem—mrem—is a measure of radioactivity; 1 mrem is estimated to increase one's risk of dying from cancer by about 1 in 4 million, corresponding to a reduction in life expectancy of about 2 minutes.) Cohen, a professor of physics and radiation health at the University of Pittsburgh, was responsible in the late 1980s for supervising the measurement of radon levels in some 350,000 U.S. homes. He puts Chernobyl's danger in context by pointing out that 16,000 deaths worldwide "is still less than the number of deaths caused every year by air pollution from coal-burning power plants in the United States" alone.

The rest of the world didn't choose to be irradiated by a badly designed and criminally misoperated Soviet nuclear power plant. Cohen's comparison is instructive but inappropriate. On the other hand, nuclear power serves useful purposes in the United States, and millions of Americans willingly buy the electricity that nuclear utilities generate. It ought to be appropriate to put nuclear-generated electricity in the context of other acceptable risks Americans take in the name of productivity, comfort, and convenience. Cohen does so, to startling effect:

> Everything we do involves risk. . . . There are dangers in every type of travel, but there are dangers in staying home—25 percent of all fatal accidents occur there. There are dangers in eating—food is one of the most important causes of cancer and of several other diseases—but most people eat more than is necessary. There are dangers in breathing—air pollution probably kills 100,000 Americans each year, inhaling radon and its decay products is estimated to kill 14,000 a year, and many diseases like influenza, measles, and whooping cough are contracted by inhaling germs. . . . There are dangers in working—12,000 Americans are killed each year in job-related accidents, and probably ten times that number die from job-related illness—but most alternatives to working are even more dangerous. There are dangers in exercising and dangers in not getting enough exercise. Risk is an unavoidable part of our everyday lives.

To quantify risk, Cohen uses the standard measurement "loss of life expectancy" (LLE), which he defines as "the average amount by

which one's life is shortened by the risk under
consideration." That's a familiar measurement;
most present and former smokers have heard
that within the context of a pack-a-day habit,
smoking one cigarette shortens a smoker's life
by about 10 minutes—in other words, has a 10-
minute LLE. Heart disease in the United States
carries an LLE of 5.8 years. Being poor in
America has catastrophic consequences for
health and well-being, carrying an LLE of about
10 years. The risk of being struck by lightning
carries an LLE of 20 hours.

Cohen puts nuclear energy in the same com-
parative context. He derives two sets of LLEs.
The first he bases on the Reactor Safety Study
issued by the U.S. Nuclear Regulatory Commission in 1975. The second, which serves as a control, he bases on a critical review of the NRC
study published in 1977 by the Union of
Concerned Scientists, an organization which has
been critical of commercial nuclear power. The
results of Cohen's LLE comparisons are startling
and reassuring:

According to the Reactor Safety Study . . . the risk
of reactor accidents [in the United States] would
reduce [each American's] life expectancy by 0.012
day, or 18 minutes, whereas the . . . Union of

Concerned Scientists (UCS) estimate is 1.5 days. Since our LLE from being killed in accidents [of any kind] is now 400 days, this risk would be increased by 0.003 percent according to the NRC, or by 0.3 percent according to the UCS. This makes [the possibility of] nuclear accidents tens of thousands of times less [a risk] than moving from the Northeast to the West (where accident rates are much higher), an action taken in the last few decades by millions of Americans, with no consideration given to the added risk. Yet nuclear accidents are what a great many people are worrying about.

Cohen then offers a Table of Loss of Life Expectancy (LLE) Due to Various Risks (the asterisks before certain activities or risks indicate averages over the total U.S. population; others refer to those directly exposed):

ACTIVITY OR RISK	LLE (days)
Living in poverty	3,500
Being male (vs. female)	2,800
Cigarettes (male)	2,300
Heart disease	2,100
Being unmarried	2,000
Being black (vs. white)	2,000
Socioeconomic status, low	1,500
Working as a coal miner	1,100

*Cancer	980
Thirty pounds overweight	900
Grade school dropout	800
*Suboptimal medical care	550
*Stroke	520
Fifteen pounds overweight	450
*All accidents	400
Vietnam army service	400
Living in Southeast (SC, MS, GA, LA, AL)	350
Mining construction (accidents only)	320
Alcohol	230
*Drug abuse	100
*Suicide	95
*Homicide	90
*Air pollution	80
Motor vehicle accidents	80
Occupational accidents	74
*AIDS	70
Small cars (vs. midsize)	60
Married to smoker	50
*Drowning	40
*Speed limit: 65 vs. 55 miles per hour	40
*Falls	39
*Poison, suffocation, asphyxiation	37

*Radon in homes	35
*Pneumonia, influenza	30
*Fire, burns	27
Coffee: 2½ cups/day	26
Radiation worker, age 18–65	25
*Firearms	11
Birth control pills	5
*All electricity, nuclear (UCS)	1.5
Peanut butter (1 tablespoon per day)	1.1
*Hurricanes, tornadoes	1
*Airline crashes	1
*Dam failures	1
Living near nuclear plant	0.4
*All electricity, nuclear (NRC)	0.04

(From Bernard Cohen, *The Nuclear Energy Option,* Plenum Press, 1990, p. 128.)

Using the above table as a guide, Cohen concludes the following:

According to the best estimates of Establishment scientists, having a large nuclear power program in the United States would give the same risk to the average American as a regular smoker indulging in one extra cigarette every 15 years, as an overweight person increasing his or her weight by 0.012 ounce, or as raising the U.S. highway speed limit from 55 to 55.006 miles per hour, and it is 2,000 times less risky than switching from midsize to small cars. If you do not trust Establishment scientists and prefer to accept the estimates of the Union of Concerned Scientists, the leading nuclear power opposition group in the United States and scientific adviser to Ralph Nader, then having all U.S. electricity nuclear would give the same risk as a regular smoker smoking one extra cigarette every 3 months, or of an overweight person increasing his weight by 0.8 of an ounce, or of raising the U.S. highway speed limit from 55 to 55.4 miles per hour, and it would still be 30 times less risky than switching from midsize to small cars.

"**N**uclear power was rejected because it was viewed as being too risky," Cohen continues, "but the best way for a person to understand a risk is to compare it with other risks with which that person is familiar. These comparisons are therefore the best way for members of the public to understand the risks of nuclear power."

Antinuclear advocates are familiar with these risk comparisons and usually counter them by arguing that "they're just estimates" (which is no longer true; they're based on thirty years of reactor experience) or that averages are misleading and "dead is dead." Cohen carries his argument one step further beyond risk comparisons to comparisons of costs per life saved and, in so doing, regains the high moral ground.

The NRC's Reactor Safety Study estimated that a nuclear power plant will cause an average of 0.8 death across its lifetime (the sum total of various risks such as potential radiation releases and accidents). Following the publication of that study, the NRC mandated safety improvements to nuclear power plants that increased their cost by $2 billion each, corresponding to $2.5 billion per life saved ($2 billion divided by 0.8). "We see here a truly horrible human tragedy," Cohen comments. "The $2.5 billion we spend to save a single life in making nuclear power safer could save many thousands of lives if spent on radon programs, cancer screening, or transportation safety. That means that many thousands of people are dying unnecessarily every year because we are spending this money in the wrong way."

Further, Cohen argues, adding to the cost of nuclear power plants made them more expensive than coal-burning plants, which utilities then ordered instead. But a coal-burning plant, says Cohen, "causes an estimated 3,000 deaths over its operating lifetime [primarily from air pollution]. Every time a coal-burning power plant is built instead of a nuclear plant, about 3,000 people are condemned to an early death—all in an attempt to save one life." (Industry sources derive a somewhat smaller number than Cohen, about 1,000.)

Hans Bethe, active in nuclear-physics research and development since its origins in the 1930s, carries the argument about the risks of nuclear power forward from estimates to facts:

In more than a quarter-century, with 112 nuclear power plants operating in the United States and an additional 200 similar ones operating throughout the world, the safety and environmental record of Western, non-Chernobyl nuclear power has been remarkable. No death or serious injury has been caused by radiation from any American-style light-water reactor anywhere in the world in thirty years of commercial nuclear power. Right now, the world needs nuclear energy sources to help reduce the massive burning of fossil fuels and their recognized environmental impact. Nuclear power is a major component of the solution to dependence on insecure and politically troublesome supplies of imported oil and also to the danger of global warming. It shouldn't be held back by inappropriate generalizations.

Evidently there's an immense gulf between scientists' and citizens' estimates of the safety of nuclear power. Whether the difference will continue to inhibit nuclear energy's further development remains to be seen. In the early 1990s it's clear that in the decades to come the United States and the world will need more electricity—a lot more electricity—and will have to find answers to the problem of global warming at the same time. The last question I want to examine is how nuclear power in the United States might improve its prospects sufficiently to meet those seemingly conflicting necessities. That question in turn takes us back to the opening pages of Chapter Two, to the second of those two historic tests I mentioned, the one that took place in Idaho in April 1986, which marked a point of renewal—a hopeful new beginning.

A SUBSTITUTE FOR HUMAN LABOR

Alan Schriesheim, the director of Illinois' Argonne National Laboratory, is a tall, friendly man. That April morning at Argonne West, he was delighted to hear the results of the crucial test his Idaho staff had just conducted. "I think you folks will look back on this day as something you'll want to tell your children about," he complimented them. They had shut off the coolant flow to a 19-megawatt experimental reactor running at full power with its emergency core-cooling system disabled. Because of its design, the advanced reactor had shut itself down safely within 100 seconds—without mechanical intervention. A related test that afternoon disabled the reactor's heat-transfer system under similar conditions, with equally successful results. "In both tests," reactor manager Peter Planchon told the press at the end of the day, "the reactor regulated its own power and temperature naturally. There was no intervention by human operators or emergency safety systems."

The reactor that Schriesheim and his colleagues were testing is a pilot model of one of a new generation of inherently safe reactors that hold great promise for the future of nuclear power. Its configuration goes back to the early years of postwar reactor design, in the tradition of Schriesheim's predecessor, Walter Zinn. It's the brainchild of the Canadian physicist Charles Till, associate laboratory director for engineering research at Argonne. I traveled to Argonne to talk to Till and find out more about the machine he named the Integral Fast Reactor (IFR). I came away impressed.

Till set out to design a reactor that would be passively safe, economically competitive, and low-waste. He saw immediately that those goals required a breeder reactor. "During the Carter era," Till told me, "I quietly put together a team to compare and study options for a breeder reactor, to see what was possible with different fuels, moderators, and configurations." The basic requirement, Till found, is that "you have to have an economical fuel cycle, so that you can burn some major percentage of your fuel. Present [light-water, slow-neutron] reactors use less than half of 1 percent of the fuel's fissile material, one time through, and then bury the lot as waste." If, instead, a reactor used fast neutrons to breed plutonium in the U-238 that constitutes the bulk of the fuel, and the operator recycled the plutonium, burn-up rates could go up to 15, 18, 20 percent, or more.

Till and his colleagues liked the properties of sodium metal as a coolant, as Zinn had liked them years before. "Sodium is a marvelous coolant," Till commented. "It has the disadvantages that it reacts with water and burns in air. Those are its only disadvantages, however. The right kind of fuel plus sodium coolant would ride right through an accident like TMI or Chernobyl. We did our 'loss-of-coolant accident' out in Idaho in 1986 and nothing happened. Nearly a month later they did that in Russia, cut off the power to the pumps, and what they got was Chernobyl."

The right kind of fuel, Till's team concluded, was metal rather than the oxide that is traditional and universal in commercial reactors today. Metal was right because it avoids the moderating effects of oxygen in the oxide, which slows and absorbs neutrons. Moderating slows down enough neutrons to allow a chain reaction to proceed, but the less moderation, the greater the number of fast neutrons and the more breeding of plutonium from U-238. Those two results lead to a more economical use of fuel over the lifetime of the reactor. Metal fuel and sodium coolant together represent the least possible moderation. Metal fuel has vastly superior thermal properties as well. "But the world went away from metal in the early 1960s," Till told me, "because metal swells when it's heavily irradiated in a reactor." Fuel elements as they were then designed consisted of uranium-alloy rods tightly wrapped with a protective cladding such as stainless steel; when the metal rods swelled, they burst their cladding and stuck in the reactor fuel-rod channels. "General Electric suggested oxide for fast reactors in the 1960s because nobody knew how to design a metal fuel that wouldn't swell," said Till.

In the late 1960s, Argonne figured out how to make workable metal-fuel assemblies. "The fuel swells," Till said, "because the radioactivity creates gases, such as argon, within its structural space, and the gas pushes against the metal. The solution was to create a fuel cartridge that surrounded the fissile metal slug with a miniature canister that contained sodium and with a plenum at the top—a space. The gas goes out into the plenum. The fuel still swells, but it reaches equilibrium at 3 percent burn-up and then burns on, and the sodium gives a perfect thermal bond."

The Integral Fast Reactor would be a pool-type reactor—the metal-fuel assembly immersed in a big tank of sodium—that would run at atmospheric pressure. "The IFR has inherent safety," Till told me, "because the flow of coolant and reactivity go in lockstep." As the Idaho tests proved, loss of coolant flow leads to heating of the metal fuel. But heating metal makes it expand. Expansion moves the atoms of uranium and plutonium in the fuel slug farther apart, which reduces reactivity. That's why the Idaho reactor passively and safely shut itself down.

A major problem with breeder-reactor prospects has been closing the fuel cycle—recycling used fuel to improve long-term efficiency. The technology used in the United States, France, and Japan to separate plutonium from waste in fuel reprocessing is called the Purex process. It was developed by the chemist Glenn Seaborg and his colleagues at Argonne's predecessor, the Metallurgical Laboratory of the University of Chicago, to produce plutonium for the first atomic bombs during World War II. It's extremely efficient, an outstanding technology for making plutonium of great purity for nuclear weapons. Unfortunately, it's also expensive, and because of the purity of its product, it carries with it the potential for nuclear proliferation. The debate at the beginning of the 1990s about Japanese plans to ship plutonium recycled in France back to Japan was founded on just this problem with the Purex process.

"Purex is too good for what you want," Till told me. "For the IFR you want U-235 as well as plutonium and the minor actinides [radioactive elements that are minor by-products of reactor operation]. The separation process needs to be simple and cheap. So Les Burris, an electrochemist here at Argonne, came in one day and proposed electrorefining." Burris, director of the chemical technology division, proposed "an electric cell, like a reverse battery," Till said. Chop the used fuel elements, dump them into a tank filled with conductive molten salts, and insert electrodes into the salts. "You set a voltage and that drags all that family of elements out in one step." The process coats a cathode with uranium, plutonium, and minor actinides in a thick, cylindrical filigree. That material in turn can be melted, mixed, cast directly into fuel elements, and reused.

Electrorefining has the added advantage of producing a radioactively "dirty" product. It can't be handled except remotely, with heavy shielding, which means no would-be terrorist can slip a bomb's worth into his briefcase and carry it home. And since it removes all the longer-lived actinides from the spent fuel, leaving only fission products behind, the process produces a much shorter-lived waste. "The radiological toxicity of the fission products in [the] waste drops below that of the original uranium ore in around two hundred years," Till and his colleague Y. I. Chang noted in a 1990 paper. Two hundred years compared with tens of thousands of years for light-water reactor waste would present a much more manageable challenge to waste-disposal technology.

Yet another advantage of the IFR is its relatively compact size. A 1,000-megawatt light-water reactor has a core 15 to 20 feet in diameter and uses fuel enriched to 3 percent U-235. A 1,000-megawatt fast reactor like the IFR would have a core with an 8-foot diameter using fuel enriched to 15 to 20 percent. The inventory of fissile material is about the same in both. The significant difference in scale, however, would mean that IFR fuel reprocessing by electro-refining could actually be built into a commercial reactor complex, eliminating the need to transport spent fuel assemblies around the country. "So the reactor could be totally self-contained," Till explained. "Put in fifty tons of uranium on the site and reuse it for fifty years. The result is then ready to bury. The process has inherent safety; the spent fuel has less potential for proliferation; you're burning it up; and there's no waste transport until the end of its life."

Although Japan and France have decided for the near future to continue conservatively with light-water reactors, converting to mixed-oxide fuel to use up their extensive Purex-generated plutonium inventories, Japan has invested $20 million in Argonne's IFR program. The U.S. Department of Energy, however, remains the principal funding source. But U.S. utilities aren't yet interested. "The nuclear power industry is hunkered down," Till told me. "They don't want to build. They want to run their present plants as cash cows."

Other new reactor designs are under development in the United States, breeder designs as well as advanced light-water systems.

And deadlines for decisions are near. Not only is electrical demand in the United States increasing, especially in areas of high population growth such as Florida, but many of the nation's first-generation commercial power reactors are approaching the end of their useful lives. Yankee Atomic Electric decided to decommission its Yankee Rowe plant in 1992 rather than spend the millions that would be required to extend its life. In the short run, utilities are finding economic advantage in building gas-turbine systems they can kick in during times of peak load. Conservation, which makes the best economic sense of all available options, continues to achieve valuable gains. But industry in the United States, as in Japan and France, is converting to electricity to reduce pollution, and per capita consumer demand is increasing as well. In the longer run—somewhere in the first decade of the 21st century—new plants will have to be built, and if they're going to be nuclear, eventually they'll have to be breeders; enough light-water reactors to generate all the world's electricity—three thousand 1,000-megawatt units—would use up the world's available uranium supply in only ten years.

Another cutoff point in reactor development is approaching that worries engineers like Stone

& Webster's Don Wille: nuclear culture in the United States is in precipitous decline. "My generation is moving toward retirement," Wille told me. "And there aren't many kids taking nuclear engineering anymore, because they don't see that it has a future. We're approaching a time when all the hands-on experience, the practical knowledge of how to make things work, is going to be gone. What happens then if the country wants to build nuclear power plants again?" The blunt, outspoken Nobel laureate physicist I. I. Rabi answered that question a few years ago, with bitter humor. "Well, when we need it," Rabi told a friend, "I guess we can buy it from the French." Or the Japanese. But to concede that possibility is to consign to foreign competition a technology in which the United States has always excelled.

The first to bring new reactors on line in the United States will probably be the Tennessee Valley Authority. Marvin Runyon, TVA chairman from 1988 until he became Postmaster General in the summer of 1992, revitalized reactor development at the huge power organization. Runyon came to TVA from the automobile industry. He had been vice-president of body and assembly operations at Ford, in charge of all North American assembly, and he had built

Nissan's famous assembly plant in Smyrna, Tennessee.

"In 1987 President Reagan asked me to become chairman of TVA," Runyon told me. "People said it would be a hell of a challenge. They asked me why I wanted to change what couldn't be changed—TVA had been increasing rates 10.4 percent per year for seventeen straight years." Four years later, Runyon is proud of his accomplishments at TVA. "TVA hasn't had a rate increase for five years now and won't have one for some years more."

Runyon expected to put nine reactors into operation by the end of 1999. "TVA shut down its reactor operations in 1985," he said. "We had bad management, bad maintenance. When I came in 1988, I stopped work on everything. My predecessors were trying to build nine reactors at the same time. You can't do that. It requires too much management, too much technical expertise. We got rid of contract managers and got out of the construction business. Two different construction companies are continuing construction of two reactors right now, and we insist on the involvement of their top management. We'd invested $8 billion in nuclear power plants that never produced one watt of power.

For less than equivalent savings from conservation, we can finish those nuclear units. That's what we're doing."

Runyon thought that TVA would need another power-generating unit by about 2004 "if demand continues to climb." That will probably be a nuclear power plant, he said. "For a brand-new unit, we'll look at conservation to see if conservation wouldn't be cheaper. But if we need to build, to me nuclear is the environmentally desirable system of choice, better than any other except hydro. We operate 11 [fossil-fueled] plants with a total of 59 coal-fired units. Coal would cost as much to build today as nuclear, and then you still have the [carbon dioxide] problem. We may build a gas peaking plant for peak-demand periods only. But when we build our next nuclear plant, we can also use it for peaking."

TVA may lead the way to renewed building of nuclear power plants in the United States, but much depends on a change in the climate of environmental politics. "What we need in this country," Runyon concluded, "is one-step licensing, and a resolution to the waste-disposal issue. It can be resolved. It has been in France and in Japan. We're handling our wastes at our own sites at the moment, and we can do that for the next four hundred years, but we need a national system."

A standardized plant is under development, and Congress has approved one-step licensing. As commercial nuclear power has continued to extend its record of safe operation in the United States, environmentalists have moved their focus to waste disposal; waste disposal is the one part of the system still under development and consequently the most vulnerable. "There's one very large problem that looms over any discussion of having two hundred or more reactors operating in this country in the next few decades," writes Susan Q. Stranahan in *Audubon* magazine. "It's a problem the industry hasn't been able to solve since the switch was first thrown at Yankee Rowe in 1961. The problem is waste disposal." Physicist Bernard Cohen counters that argument impatiently:

Disposal of high-level [nuclear] waste is often referred to as an "unsolved problem." When people say this to me, I ask whether disposal of the waste from coal burning—releasing it as air pollution, killing about 30,000 Americans each year—is a "solved problem." . . . When I point out that the radioactive-waste problem is not unsolved since there are many known solutions, the usual retort is

"Then why aren't we burying it now?" The answer to that question is that we are in the process of choosing the best solution. . . .

I like to point out that nobody is being injured by the high-level waste, and there is no reason to believe that anyone will be injured by it in the foreseeable future. . . . The usual reply is "How do you know that nobody is being injured?" The answer is that we constantly measure radiation doses in and around nuclear facilities and throughout our environment, and nobody is receiving any such doses from high-level waste. If there is no radiation dose, there can be no harm.

Cohen concludes in words that would settle the issue in any context less emotional and irrational than nuclear power: "Having studied this problem as one of my principal research specialties over the past fifteen years, I am thoroughly convinced that radioactive waste from nuclear power operations represents less of a health hazard than waste from any other large technological industry."

But the convictions of scientists, however well supported by evidence, are not, of course, the crux of the issue in a democratic society. The nuclear industry still needs to deal with the convictions of worried citizens as well. So far, it has avoided doing so, preferring to concentrate instead on promoting new federal legislation to streamline licensing (which environmentalists charge is intended to reduce public input) and on national advertising that promotes nuclear power as antipollution and antigreenhouse. The operant principle seems to be to let sleeping dogs lie. The Japanese and the French, with their on-site science centers and lectures to schoolchildren, know better. In the long term, democratic societies work by consensus, not by authority. Authority gave the world Chernobyl.

An aggressive program to educate the public about nuclear waste would distinguish the continuing success of civilian nuclear-waste management from the appalling mess that half a century of weapons production has left behind. It would compare waste from fossil-fuel plants with waste from nuclear plants and assess their very different degrees of relative risk. A large coal-burning plant, Cohen notes, produces 15 tons of carbon dioxide a minute, as much nitrogen oxide as 200,000 automobiles, several pounds of particulates per second, 1,000 pounds of ash per minute, enough organic carcinogens to cause cancer deaths in two or three people per year. Radon gas generated from radium in coal, released when coal is burned, adversely affects the public's health more than all the radioactive waste released from nuclear power plants; the waste from a comparable nuclear plant during one year of operation is also about five million times smaller by weight and billions of times smaller by volume, weighing about one and a half tons and occupying a volume of half a cubic yard, a quantity so small that "it can be handled with a care and sophistication that is completely out of the question for the millions of tons of waste spewed out annually from our analogous coal-burning plant," says Cohen.

Siting nuclear-waste disposal facilities continues to be politically volatile. Herbert Inhaber, of Ecology and Environment Inc. in Lancaster, New York, has proposed a workable, market-based approach that deserves industry and government consideration: a reverse Dutch auction. "A Dutch auction," Inhaber writes, "has a falling price—sometimes set on a clock—and only one bid. When the bidder signals, the auction is over. . . . Dutch auctions deal with desirable objects. In the case of unwanted and undesirable land use, the auction must run in reverse." Inhaber sees a reverse Dutch auction as a way for citizens to gain control through economic mechanisms over nuclear- or other toxic-waste siting. If people resist technological change primarily because they feel they are being denied choice and railroaded, as many studies have shown, then Inhaber's system might give them the choice they feel they lack, and they would feel that choice in their pocketbooks, where it counts. "The principle," he notes, "is to encourage a local community to volunteer to accept a waste site," building consensus rather than imposing authority. Here's how it might work:

The first step in the reverse Dutch auction is the announcement of the environmental or risk ground rules, including a maximum population density, ground-water considerations, geological regulations, endangered-species rules, and so on. Because no specific local region has been identified at the beginning of the process, these ground rules can be chosen relatively objectively. . . .

When the auction begins, the siting authority publicly offers any volunteer region an amount of money if it can find an environmentally acceptable site. The bonus keeps rising steadily, to entice regions that had hitherto sat on the sidelines to select a site. Eventually, one region will suggest a site when the true cost of the facility—the price the rest of us have to pay to have the site away from our back yard—is reached.

Inhaber notes that this sensible system is analogous to the system that was used to determine commercial reactor sites in the first place: "The NRC has never chosen a reactor site; this was always done by a utility, which presumably had a reason for building the facility. The NRC approved or disapproved the site only after submission by the utility." Nor, he argues, would the reverse Dutch auction be a bribe since it's done publicly, it isn't targeted to any specific organization, and it wouldn't be illegal. "The reverse Dutch auction is an example of the use of an incentive, rather than a bribe, to solve a difficult problem." Unlimited liability—the possibility that the price might rise beyond what the rest of the nation would wish to pay—could be solved by setting a cap on the bonus to be offered. Finally, the reverse Dutch auction is, in Inhaber's words, "market- rather than bureaucracy-driven," and therefore "removes the alleged coercion that has marked most attempts to site these facilities."

Lawrence Lidsky, professor of nuclear engineering at the Massachusetts Institute of Technology, has offered another proposal that might increase public confidence in commercial nuclear power: licensing new designs by test, as new commercial aircraft designs are licensed. Lidsky told a conference at MIT in 1990, "I believe that the development of nuclear power plants with the ability to survive a definitive worst-case, 'absolute' test is a minimum requirement if nuclear power is to play a significant role in the future. The test protocols . . . include, at a minimum, simultaneous loss of coolant, control-rod withdrawal, and the presence of a malicious operator."

If a new reactor design passed such a test, Lidsky pointed out, "there would be no need for detailed regulatory oversight nor justification for rule changes in plants identical to the tested plant. The role of the NRC would change to that of inspectors at the vendor's facility, ensuring that the plants were being manufactured to original specifications. . . . The NRC would evolve to a Federal Aviation Administration–like organization. . . . This is a market-based rather than a regulatory-based approach to standardization."

I asked Charles Till at Argonne if he thought the IFR could pass such a rigorous test. He said he thought it could.

Unlike France and Japan, the United States has alternatives to nuclear power. That's the primary and essential reason why nuclear power stalled when the cost of construction went too high. The United States has several hundred years' worth of coal reserves and at least twenty years' worth of natural gas. There's room for major improvement in conservation as well, particularly in domestic use of electricity.

But coal is highly polluting. So is oil. Natural gas is notably less so but adds a full burden of carbon dioxide to the greenhouse effect, a disadvantage that is giving rise to increasingly serious concern as its consequences begin to be felt throughout the world. Natural gas is probably the best near-term solution to automobile pollution, but even automobiles will need to be shifted to electrical power in the decades to come, a process California is already spearheading. Indeed, as the French experience makes clear, electricity from nuclear power is the best of all solutions (after conservation) to both pollution and greenhouse warming. With breeder technology like Argonne's Integral Fast Reactor, it's also the best—in fact, the only—present answer to the long-term energy needs of the world. Conservation and stopgap measures may serve for the first decade or so of the 21st century, but what about the next hundred years?

Whether nuclear power will survive to make a contribution depends, first of all, on whether the West succeeds in helping Central Europe, Russia, and Ukraine improve the safety of their ailing nuclear reactors—their legacy from the secretive, authoritarian system that built them so badly and operated them so unsafely that they strike terror in the hearts of nuclear operators throughout the industrial world. In Bulgaria, in Czechoslovakia, in Mongolia, and most of all, in the countries of the former Soviet Union, reactors as badly protected and as carelessly operated as Chernobyl clunk on—accidents waiting to happen. Nuclear operators in other countries have moved quickly to help, but national pride and lack of money and of alternative sources of electricity have limited solutions.

Assuming that another Soviet-made reactor doesn't blow, the U.S. nuclear industry still has to convince Americans that nuclear power and waste disposal are safe. Avoiding the problem or only paying it lip service won't make it go away. The chasm between utility executives and the public remains far too wide. One senior engineer I interviewed pointed to an electrical outlet in his office when I mentioned public distrust. "See that plug over there?" he asked me. "That's where most people think electricity comes from. When the blackouts start, people won't know or care how their electricity is made." After all the trouble the nuclear power industry brought down on itself by underestimating the democratic process and the good judgment of the American people, such cynicism is remarkable.

Most of all, nuclear power needs to be seen in context. It pollutes less than any other major source of energy. It supplies clean, efficient electricity at competitive rates. It's a long-term source. It has the potential to solve other problems such as automobile pollution and dependence on foreign oil. Rather than an option, it's almost certainly going to be a necessity.

David Lilienthal, the first chairman of TVA and of the Atomic Energy Commission, came to similar conclusions when he discussed the energy future in his book *Atomic Energy: A New Start*:

We rely heavily on nuclear power to keep our economy going. . . . For the near- and long-term future, the energy we now have and can count on, from all sources, is not enough. Except for temporary periods, it has never been enough, and it will never be enough for the kind of developing country we are, with our population steadily increasing and our desires and incomes expanding without long-term letup. I have listened for years to assertions that we don't need more energy; they have always and everywhere been wrong, and they are just as wrong today as they have been throughout the history of energy and industrialized economies.

"Energy is part of a historic process," Lilienthal concludes wisely, "a substitute for the labor of human beings. As human aspirations develop, so does the demand for and use of energy grow and develop. This is the basic lesson of history."

Satisfying human aspirations is what our species invents technology to do. Some Americans, secure in comfortable affluence, may dream of a simpler and smaller world. However noble such a dream appears to be, its hidden agenda is elitist, selfish, and violent. Millions of children die every year for lack of adequate resources—clean water, food, medical care—and the development of those resources is directly dependent on energy supplies. The real world of real human beings needs more energy. With nuclear power, that energy can be generated cleanly and without destructive global warming. Whether it will be or not depends on leadership and public education. Where nuclear power is concerned, in both departments the United States has a long way to go.

F O R
F U R T H E R
R E A D I N G

Hans Bethe. "The Necessity of Fission Power." *The Road from Los Alamos.* Touchstone, 1991.

Gail Bingham and Daniel S. Miller. "Prospects for Resolving Hazardous-Waste Siting Disputes Through Negotiation." *Natural Resources Lawyer* 17:473 (1984).

Bernard L. Cohen. *The Nuclear Energy Option.* Plenum Press, 1990.

James M. Jasper. *Nuclear Politics.* Princeton University Press, 1990.

Grigori Medvedev. *The Truth About Chernobyl.* Basic Books, 1991.

Paul Slovic et al. "Perceived Risk, Trust ,and the Politics of Nuclear Waste." *Science* 254:1603 (13 December 1991).